D0140477

Applications of Abstract Algebra

George Mackiw

WAGGONER LIBRARY
DISCARD

John Wiley & Sons
New York / Chichester / Brisbane / Toronto / Singapore

MACKEY LIBRARY
TREVECCA NAZARENE COLLEGI

Copyright ©1985 by John Wiley & Sons, Inc.

All rights reserved.

Reproduction or translation of any part of
this work beyond that permitted by Section
107 or 108 of the 1976 United States Copyright
Act without the permission of the copyright
owner is unlawful. Requests for permission
or further information should be addressed to
the Permissions Department, John Wiley & Sons, Inc.

ISBN 0-471-81078-9
Printed in the United States of America

10 9 8 7 6 5 4 3 2 1

Preface

Abstract or modern algebra is considered by many to be one of the more beautiful, elegant and coherent branches of mathematics. Yet, in spite of the greater emphasis today on the use of discrete methods, the study of groups, rings and fields seems to have suffered. A primary reason may be that traditional presentations of the subject have not placed much emphasis on motivation and applications, providing perhaps less reason for the average student to be attracted to or to study the subject. For example, while it is often mentioned that symmetry groups are important in chemistry and physics, rarely are concrete examples of such groups given in introductory texts.

The goal of this text is to provide a collection of applications of abstract algebra to a student who is currently taking a one or two semester introduction to the basics of groups, rings and fields. The text is thus intended for use as a supplement to an Abstract Algebra course, though it certainly may also be used in a separate course on applications or as a reference. The applications included here cover exact computing, error-correcting codes, the construction of block designs useful in statistics, crystallography, integer programming, cryptography and combinatorics. A more detailed description of individual chapters follows below.

My motivation in writing this text was to present a series of applications that would be accessible to the student in a first course, provide examples of algebra in use, contain topics not previously gathered together in text form, and hopefully be interesting. Topics such as classical Greek constructions and applications of Galois Theory have deliberately been omitted since these are precisely the ones that are covered in many excellent texts. Applications to quantum mechanics and physics are also absent since they tend to require a background deeper than that possessed by the beginning student.

The chapters that follow are, with two exceptions, independent of each other. We include a description of the content and prerequisites for each chapter.

1) Computing in Z_M - Exact Solutions of Linear Equations

Problems of round-off and precision are inherent in the floating-point arithmetic used in computers. This chapter describes how the arithmetic of the ring Z_M and the Chinese Remainder Theorem may be helpful in avoiding such problems when dealing with systems of linear equations.
Prerequisites: elementary linear algebra, modular arithmetic, ring homomorphisms.

2) Block Designs

The use of abelian groups, finite fields and rings in the construction of block designs, structures which are useful in the design of experiments.

Prerequisites: elementary theory of abelian groups, modular arithmetic, the fields $GF(p^n)$, primitive elements.

3) Error Correcting Codes I: Hamming Codes

An introduction to the basics of error correcting codes. This chapter features a detailed development of the Hamming codes and the rationale behind their error correcting capabilities.
Prerequisites: elementary linear algebra, the field $GF(2)$, cosets, the fundamental theorem of group homomorphisms.

4) Error Correcting Codes II: BCH Codes

This requires the previous chapter and features the development of the multiple error correcting BCH codes. It includes a step by step exposition of the error correcting procedures attached to BCH codes, which serve as a wonderful primer for reinforcing the arithmetic of finite fields.
Prerequisites: elementary linear algebra, the finite fields $GF(2^n)$, primitive elements, minimal polynominals.

5) Crystallographic Groups in the Plane I

This serves as an introduction to the classification of crystals via their symmetry groups. Introduces the notion of crystallographic space groups, translation subgroups and point groups in two dimensions.
Prerequisites: elementary linear algebra and group theory, quotient groups, fundamental theorem of group homomorphisms.

6) Appendix: Orthogonal Matrices in Two Dimensions

Develops from scratch results on 2 x 2 orthogonal matrices which are needed in the above chapter.
Prerequisites: familiarity with 2 x 2 matrices and determinants.

7) Crystallographic Groups in the Plane II

This is a continuation of the previous chapter and provides a rigorous development of the fact that there are only 17 inequivalent crystallographic groups in the plane. These are illustrated by concrete matrix representations and drawings.
Prerequisites: same as the previous chapter.

8) The RSA Public Key Cryptosystem

Discusses the issues behind modern public key cryptography and develops the algebra underlying the Rivest-Shamir-Adleman system. Includes a discussion of recent developments in primality testing and factorization.
Prerequisites: modular arithmetic, the ring Z_M, Lagrange's Theorem for groups.

9) Integer Programming

        Introduces integer programming through optimization
problems. Develops the role finite abelian groups play in the solution
of integer programming problems.
Prerequisites: elementary linear algebra and linear programming, use of
the simplex algorithm, group homomorphisms, quotient groups, the
exponent of a finite abelian group.

10) Group Theory and Counting

        The role group theory plays in combinatorial problems;
includes a derivation of Burnside's Lemma with many concrete examples.
Prerequisites: elementary group theory, cosets, equivalence relations.

        As may be evident, elementary linear algebra is invoked
throughout. This should present no significant difficulty, since it is
now commonplace for most students to encounter linear algebra in either
freshman or sophomore year. With the exception of the two chapters each
on crystallography and error correcting codes, the individual sections
of the text are independent of each other and may be used in any order
desired. Each chapter includes a set of exercises and a list of further
references.

        I should like to thank the staff of the Loyola College
Communication Center, especially Charlette Hinchliffe, Nancy Marshall,
Mary Muhler, and Marion Weilgosz for their diligent work in preparing
this manuscript. Dean David Roswell was very helpful in his support and
encouragement. It was my pleasure to engage in many interesting
conversations with Professor Alan Goldman of Johns Hopkins University
during a sabbatical year spent at the Center for Applied Mathematics at
the National Bureau of Standards. It was he who provided the idea for
the chapter on integer programming. It is my hope that both students
and teachers of abstract algebra will find this collection useful.

                                        George Mackiw
                                        Baltimore, Maryland

Contents

Contents

1 Computing In Z_M: Exact Solutions of Linear Equations

The very nature of computer arithmetic carries with it the possibility of error propagation due to, say, truncation, round-off or the inherent sensitivity of the problem under consideration. If, for example, the system

$$(1.1) \quad \begin{aligned} 1.0000x_1 + 2.0000x_2 &= 5.0000 \\ 1.0000x_1 + 2.0005x_2 &= 5.0010 \end{aligned}$$

were read by a machine as

$$1.000x_1 + 2.000x_2 = 5.000$$

$$1.000x_1 + 2.001x_2 = 5.001$$

the desired solution of $x_1 = 1$, $x_2 = 2$ would instead be converted to $x_1 = 3$, $x_2 = 1$. The ability to do arithmetic on a computer 'exactly', with no tolerance for error, can obviously be of great advantage, especially in investigations where the very form of an answer is more crucial than any numerical approximations. In this section we describe an algebraic technique due to Morris Newman [8] which produces exact solutions to linear systems with integer coefficients. The basic idea is to consider such systems over Z_M, instead of Z; solve them exactly in Z_M, for perhaps several values of M, and then pull back via the Chinese Remainder Theorem to the desired exact solution. The restriction to systems with integer coefficients is not serious since any system entered in a fixed word length machine is a system with rational coefficients which can be converted to an equivalent integral system by scaling. Thus, (1.1) is equivalent to

$$10000x_1 + 20000x_2 = 50000$$

$$10000x_1 + 20005x_2 = 50010$$

A Simple Overview

For a fixed modulus $M \geq 2$, we denote by $\bar{x}$ the congruence class of the integer x, mod M.

The ring Z_M is then given by $Z_M = \{\bar{x} \mid x \in Z\}$. Note that the order of Z_M is M and that for each x there is a unique integer i, $0 \leq i \leq M-1$, with $\bar{x} = \bar{i}$.

The mapping $\phi : Z \longrightarrow Z_M$ by

$$\phi : x \longrightarrow \overline{x}$$

is a homomorphism of rings whose kernel is the ideal in Z consisting of all multiples of M. The fact that ϕ preserves addition and multiplication allows us to push computations from Z to Z_M. We illustrate this fundamental fact by a simplistic <u>example</u>:

Suppose we were concerned with evaluating the expression $w = xy - z$ for $x = 10$, $y = 29$ and $z = 300$. Neglecting the fact that most of us feel eminently competent to handle this problem, we proceed with a roundabout discussion of it.

Using the modulus M = 17, we have

$$\overline{w} = \overline{xy - z}$$
$$= \overline{x}\,\overline{y} - \overline{z}$$

and, corresponding to the indicated values of the arguments,

$$\overline{w} = \overline{10}\ \ \overline{29} - \overline{300} = \overline{10}\ \ \overline{12} - \overline{11}$$
$$= \overline{120} - \overline{11} = \overline{1} - \overline{11} = \overline{-10} = \overline{7}.$$

At this stage we then know that w must be a number whose remainder is 7 upon division by 17 --- w is thus in the set

(1.2) $\{\ldots, -10, 7, 24, \ldots\}$.

Some added information could narrow the range further. Knowing that $-30 \leq w < 10$, would limit the possibilities to $w = 7$, -10, or -27. With a priori knowledge that $-17 < w \leq 0$, we could assert that $w = -10$.

We could have also computed $\overline{w}$ for another modulus, say M = 13. Then

$$\overline{w} = \overline{10}\ \ \overline{29} - \overline{300} = \overline{10}\ \ \overline{3} - \overline{1} = \overline{29} = \overline{3}.$$

Thus w is in the set

(1.3) $\{\ldots, -23, -10, 3, 16, 29, \ldots\}$

and hence in the intersection of the sets

(1.2) and (1.3): $\{\ldots, -231, -10, 211, \ldots\}$.

Knowing ahead of time, say, that $|w| < 200$ would then also lead us to the conclusion $w = -10$.

Some naive observations:

2

1) Intuitively, arithmetic in Z_M is easier than arithmetic in Z.

2) Application of ϕ (the barring operation, "——") pushes us into Z_M.

3) Solving the problem in Z_M gives us a clue as to what the Z-solution ought to be.

4) Having an a priori estimate of the size of the solution can further narrow the range.

5) Working with several distinct M's can induce a finer sifting.

Systems of Equations

The problem we consider is that of finding an <u>exact</u> solution to the system of equations

(1.4) $$Ax = b$$

where A is an n x n invertible matrix with integer entries and b is an n x 1 column vector with integer entries.

Our first step is to pick an M and reduce everything in sight (mod M). We will use the following <u>remarks</u>:

1) By $\overline{A}$ we mean the matrix with coefficients in Z_M whose entries are 'the bars' of the entries of A. $\overline{b}$ is taken likewise.

2) Note that $\det\overline{A} = \overline{\det A}$. This is due to the fact that ϕ is a ring homomorphism and that detA is a sum of products of entries of A.

3) Since A is invertible, d = detA satisfies d $\neq$ 0. It may happen, though, that $\overline{d}$ = 0 .
   (Example: M = 3, $A = \begin{pmatrix} 2 & 1 \\ -1 & 1 \end{pmatrix}$)

4) When is $\overline{A}$ invertible as a matrix with entries in Z_M?
   Answer: precisely when d = detA satisfies the gcd condition (d,M) = 1; put another way, precisely when $\overline{d}$ is a unit in the ring Z_M. The reason for this lies in the fact that the usual results involving invertibility and determinants carry over to commutative rings, with units playing the role of non-zero field elements (see Exercise 5).

Given (1.4), we find an M with (M,d) = 1 and consider the system

3

$$(1.5) \qquad \overline{A}x^* = \overline{b} \qquad\qquad\qquad \text{over } Z_M.$$

(We will postpone questions relating to ways of choosing M until the end).

Example

Suppose the original system is

$$7x_1 + 3x_2 = 5$$
$$2x_1 + 3x_2 = 1$$

Here, d = 15. For M = 7

$$\overline{A} = \begin{pmatrix} 0 & 3 \\ 2 & 3 \end{pmatrix}, \qquad \overline{b} = \begin{pmatrix} 5 \\ 1 \end{pmatrix}.$$

For M = 4

$$\overline{A} = \begin{pmatrix} 3 & 3 \\ 2 & 3 \end{pmatrix}, \qquad \overline{b} = \begin{pmatrix} 1 \\ 1 \end{pmatrix}$$

Note: For ease of reading, we do not include 'bars' when writing elements of Z_M in examples; one need only remember that arithmetic is being done modulo M.

Since $\overline{A}$ is invertible over Z_M, the system (1.5) has the unique solution

$$x^* = \overline{A}^{-1} \overline{b} .$$

In practice, Gaussian (or Gauss-Jordan) elimination can then be used to find x*.

(1.6) Example

Consider again

$$7x_1 + 3x_2 = 5$$
$$2x_1 + 3x_2 = 1$$

with M = 31.

We then solve the system

$$\begin{pmatrix} 7 & 3 & | & 5 \\ 2 & 3 & | & 1 \end{pmatrix} \qquad \text{over } Z_{31} \text{ using}$$

elementary row operations.

$$\begin{pmatrix} 7 & 3 & | & 5 \\ 2 & 3 & | & 1 \end{pmatrix} \quad \begin{matrix} 3 \text{ times} \\ \text{second row} \\ \sim \\ \text{from first} \end{matrix} \quad \begin{pmatrix} 1 & -6 & | & 2 \\ 2 & 3 & | & 1 \end{pmatrix}$$

$$\begin{matrix} \text{twice first} \\ \sim \\ \text{row from} \\ \text{second} \end{matrix} \begin{pmatrix} 1 & -6 & | & 2 \\ 0 & 15 & | & -3 \end{pmatrix} \quad \begin{matrix} \text{multiply} \\ \sim \\ \text{second row} \\ \text{by } -2 \end{matrix} \begin{pmatrix} 1 & -6 & | & 2 \\ 0 & 1 & | & 6 \end{pmatrix}$$

$$\begin{matrix} \text{add } 6 \\ \sim \\ \text{times second} \\ \text{row to first} \end{matrix} \begin{pmatrix} 1 & 0 & | & 7 \\ 0 & 1 & | & 6 \end{pmatrix}$$

Thus, $x^* = \begin{pmatrix} 7 \\ 6 \end{pmatrix}$

Remarks:

1. If, as in the above example, M is a prime, then Z_M is a field. It should then be clear that the Gaussian elimination one is used to applying, say, to real matrices is also valid in this case.

2. It is interesting to note that, even when M is not prime (and hence Z_M contains zero divisors), the assumption that $\overline{A}$ is invertible still allows the use of row-reduction to compute x*. This is the content of Exercise 7.

The next question we pose is: what is the relation between x* and the desired solution, x, of Ax = b? (Careful: x is, in general, a vector with <u>rational</u> coefficients. It may make no sense to speak of $\overline{x}$, let alone assert that $\overline{x} = x^*$).

We recall some facts from matrix algebra. The <u>adjoint</u> of A = (a_{ij}), written adjA, is the n x n matrix whose $(i,j)^{th}$ entry is the $(j,i)^{th}$ co-factor of a_{ij}.

i.e. adjA $=$ (b_{ij}), where $b_{ij} = (-1)^{i+j} M_{ji}$

and M_{ji} is the determinant of the (n - 1) x (n - 1) matrix formed by omitting the jth row and i^{th} column of A. Note that if A has integer entries, then so does adjA.

A basic result from matrix theory states that

5

(1.7) $A(adjA) = dI$, where $d = detA$.

So, using the A and b of (1.4), we have

$A(adjA)b = db$.

Let $y = (adjA)b$. Note that y is a vector with integer entries.

Thus, $Ay = db$

and $\overline{A}\,\overline{y} = \overline{d}\,\overline{b}$

But, we know that $\overline{A}x^* = \overline{b}$.

So also $\overline{A}(\overline{d}x^*) = \overline{d}\,\overline{b}$.

Since $\overline{A}$ is invertible over Z_M, it follows that

$$\overline{y} = \overline{d}x^* .$$

<u>Remarks:</u>

1. Knowledge of $\overline{d}$ and x^* will then produce $\overline{y}$.

2. Assume for a moment that d and y can be recovered from $\overline{d}$ and $\overline{y}$, respectively.

    From (1.7) we have that

    $$A^{-1} = \frac{1}{d}\, adjA$$

    Thus, $x = A^{-1}b = \frac{1}{d}(adjAb)$

    $$= \frac{1}{d}\, y .$$

    That is, we could then recover x.

3) Write $y = \begin{pmatrix} y_1 \\ y_2 \end{pmatrix}$, where y_1, y_2 are integers.
    Let $|y| = max(|y_1|, |y_2|)$.

    If we knew ahead of time that $|d| < \frac{M}{2}$ and $|y| < \frac{M}{2}$ then each of the integers d, y_1, y_2 would lie in the open interval $\left(\frac{-M}{2}, \frac{M}{2}\right)$.

    Having computed $\overline{d}$ and $\overline{y}$, we could then identify d and y, since no two distinct intergers in $\left(\frac{-M}{2}, \frac{M}{2}\right)$ are congruent (mod M).

4) Thus, we choose the modulus M with

(1.8) $M > 2\, max(|d|, |y|)$.

Let us illustrate our comments by going back to a previous

6

(1.9)   Example:

We had considered the system

$$7x_1 + 3x_2 = 5$$

$$2x_1 + 3x_2 = 1$$

over Z_M with $M = 31$.

Here, $d = 15$, and

$$y = (adjA)\ b = \begin{pmatrix} 3 & -3 \\ -2 & 7 \end{pmatrix} \begin{pmatrix} 5 \\ 1 \end{pmatrix}$$

$$= \begin{pmatrix} 12 \\ -3 \end{pmatrix} \quad .$$

So, $|y| = 12$,

and it is true that

$$M > 2\ max(15,12).$$

It is important to note that we are working with an easy example where y and d are readily available. In practice, neither would be easily computable -- if so, the problem would already be solved. The serious question of how to choose M without first computing y and d will be addressed shortly.

The method then proceeds as follows:

1) Compute x*:

we have already shown $x* = \begin{pmatrix} 7 \\ 6 \end{pmatrix}$.

2) Compute $\overline{d}$:

Here, $\overline{d} = 15$. Again we note, that unlike this example, in practice we would wish to compute $\overline{d}$ without incurring the expense of computing d.

3) Find d:

Write $\overline{d} = \overline{d'}$, where d' is in $\left(\frac{-M}{2}, \frac{M}{2} \right)$. Then, since $|d| < \frac{M}{2}$, we have d = d'. Here, this gives d = 15.

4) Then $\overline{y} = \overline{d}x* = 15 \begin{pmatrix} 7 \\ 6 \end{pmatrix}$.

$$= \begin{pmatrix} 105 \\ 90 \end{pmatrix} = \begin{pmatrix} 12 \\ 28 \end{pmatrix}$$

7

$$= \begin{pmatrix} 12 \\ -3 \end{pmatrix}.$$

Recall that we are working over Z_{31}. Also, the integers in the last vector have been deliberately chosen to lie in $\left(\frac{-M}{2}, \frac{M}{2} \right)$.

5) Find y:

Since $|y| < \frac{M}{2}$, it follows that $y = \begin{pmatrix} 12 \\ -3 \end{pmatrix}$.

6) Find x:

$$x = \frac{1}{d} y = \frac{1}{15} \begin{pmatrix} 12 \\ -3 \end{pmatrix}$$

$$= \begin{pmatrix} \frac{12}{15} \\ \frac{-3}{15} \end{pmatrix} = \begin{pmatrix} \frac{4}{5} \\ \frac{-1}{5} \end{pmatrix}.$$

Comments:

1. The entire algorithm is conducted using only the residue arithmetic of Z_M with the exception of the computation of

   $x = (1/d)y$ at the very end. Even then, x can be formally printed in rational form to avoid any round-off. Thus, if M is such that the computer can add and multiply any two integers of absolute value less than M, the necessary arithmetic can be carried out exactly.

2. The computation of $\overline{d}$ need not involve previous computation of d. The point is that $\overline{d}$ can be computed while applying Gauss-Jordan elimination to $(\overline{A} \mid \overline{b})$. The only row operations that change the determinant of a matrix are

      1) interchanging two rows --
         in which case the determinant changes by a minus sign

   and

      2) multiplying a row by the inverse of the element c (c is then called a pivot) -- in which case the new matrix has determinant c^{-1} times the determinant of the old matrix.

   These remarks can be used to show that then

   $$\overline{d} = \det \overline{A} = (-1)^S \times \Pi \, c_i$$

   where s is the number of row interchanges, and the c_i are all the pivots used. Thus, from the work of Example (1.6), we see that $\overline{d} = (-1)^0 \, 15 = 15$.

3. In practice M is often chosen to be a prime, so that Z_M is

a field. The Euclidean algorithm can then be used to compute inverses, when necessary, in Gaussian elimination.

4. The size of M:

For $A = (a_{ij})$, $b = (b_i)$, let K be the largest of the numbers $|a_{ij}|$, $i,j = 1,\ldots,n$ and $|b_k|$, $k = 1,\ldots n$.

A simple estimate using the definition of determinant shows that

$$|d| \leq n!K^n.$$

Likewise, since $y = A^{adj}b$, it also follows that

$$|y| \leq n!K^n.$$

Thus, choosing M to be a prime bigger than $2n!K^n$, gives us the desired

$$M > 2 \max(|d|,|y|).$$

The problem is that the above estimates are crude and in many cases unnecessarily conservative. In our example (K = 7, n = 2), we would be picking a prime larger than $2 \cdot 2! \cdot 7^2 = 196$; yet M = 31 does the job. Howell and Gregory [5] discuss other estimates for M based on Hadamard's inequality for $|detA|$.

5. If M is a prime with $M > 2|d|$, then the requirement that (M,d) = 1 is clearly met.

Using several moduli: The Chinese Remainder Theorem

The whole purpose of working modM is to avoid arithmetic operations that would lead to overflow and round-off error. But to achieve this the modulus M must meet the requirement $M > 2\max(|d|,|y|)$ and this in itself can create a problem if, say, M exceeds the wordsize of the computer. A way out of this is to work simultaneously over several moduli m_i, where $M = \pi m_i$. Since the heart of this variation rests on an old and famous result called the Chinese Remainder Theorem (CRT), we examine this theorem from several points of view.

An abstract version:

Suppose the positive integer M is written as $M = m_1 m_2 \cdots m_k$,

9

where $(m_i, m_j) = 1$, if $i \neq j$.

Consider the mapping:

$$\lambda: Z_m \longrightarrow Z_{m_1} \times Z_{m_2} \times \ldots \times Z_{m_K}$$

given by

$$\lambda(\overline{x}) = (\overline{x}^1, \overline{x}^2, \ldots, \overline{x}^k).$$

Here, $\overline{x}^i$ refers to the congruence class of x (mod m_i).

It is a straightforward check to show that λ is a well-defined homomorphism of rings. Further, λ is injective (1 - 1), since if $\overline{x}^i = \overline{0}$, $\forall$ i, then x is divisible by all the m_i; hence x is divisible by their least common multiple M, so that $\overline{x} = \overline{0}$. Since the rings Z_M and $Z_{m_1} \times Z_{m_2} \times \ldots \times Z_{m_k}$ have the same order, it follows that λ is also onto -- i.e., the two rings are isomorphic.

As a consequence of this, given residue classes $\overline{a}_i$, i = 1,...,k, in Z_{m_i}, there then exists a unique class $\overline{x}$ in Z_M with $\lambda(\overline{x}) = (\overline{a}_1, \overline{a}_2, \ldots, \overline{a}_k)$.

Put another way, we have shown that the simultaneous congruences

(*)
$$
\begin{aligned}
x &\equiv a_1 \pmod{m_1} \\
x &\equiv a_2 \pmod{m_2} \\
&\;\vdots \\
x &\equiv a_k \pmod{m_k}
\end{aligned}
$$
where $(m_i, m_j) = 1$, if $i \neq j$.

always have a solution x_0, and that all solutions x' of (*) satisfy $x' \equiv x_0 \pmod{M}$, $M = m_1 m_2 \cdots m_k$

The problem is our proof is in essence an existence proof. It rests primarily on what is called the pigeonhole principle and gives no way of actually producing such an x_0. A proof that embodied a constructive procedure for finding a solution to (*) would be more satisfying to the soul and certainly more useful in applications. We thus revisit the CRT with a more computational mindset.

A constructive version:

Consider again the situation in (*). With $M = m_1 m_2 \cdots m_k$, let

$$M_i = \frac{M}{m_i}, \quad i = 1, \ldots, k.$$

Since the m_i's are pairwise relatively prime, we have that $(M_i, m_i) = 1$, for $i = 1, \ldots, k$.

It then follows that there exists an integer c_i with $c_i M_i \equiv 1 \pmod{m_i}$. Also, c_i can be found via the Euclidean algorithm (cf. Exercise 6).

Construct

$$x_0 = a_1 c_1 M_1 + a_2 c_2 M_2 + \ldots + a_k c_k M_k.$$

Since $M_j = 0 \pmod{m_i}$, if $i \neq j$, and $a_i c_i M_i \equiv a_i \pmod{m_i}$ by choice of c_i, we thus have that

$$x_0 \equiv a_i \pmod{m_i}, \quad \forall \, i \, .$$

That is, we have produced a solution to (*).

A direct verification that all solutions x' to (*) satisfy $x' \equiv x_0 \pmod{M}$ is left to the reader.

An algorithm

We can thus handle our original problem of solving $Ax = b$ exactly by using an M with $M = m_1 m_2 \cdots m_k$, doing computations in the Z_{m_i} and then using the CRT to represent our answer in Z_M. The algorithm would proceed as follows.

To solve $Ax = b$,

1) Choose an $M > 2 \max (|d|, |y|)$.

2) Write $M = m_1 m_2 \cdots m_k$, where $(m_i, m_j) = 1$, if $i \neq j$.

3) Consider the k systems

$$\overline{A}x_i^* = \overline{b} \text{ in } Z_{m_i} \, , \quad i = 1, \ldots, k.$$

4) As in the case of a single modulus, find solutions, x_i^*.

5) Also, find $\overline{d}$ for each of the m_i: call these values $\overline{d}_i$.

6) Use the CRT to solve the simultaneous congruences

$$
\begin{aligned}
x &\equiv x_1^* \pmod{m_1} & \qquad x &\equiv d_1 \pmod{m_1} \\
&\quad\vdots & &\quad\vdots \\
x &\equiv x_k^* \pmod{m_k} & \qquad x &\equiv d_k \pmod{m_k}.
\end{aligned}
$$

Call the solutions to these congruences x^* and $\overline{d}$.

11

7) Let $\bar{y} = \bar{d}x^*$.

8) Write y so that $|y| < \dfrac{M}{2}$

   and d so that $|d| < \dfrac{M}{2}$.

9) Then $x = \dfrac{1}{d} y$ is a solution .

We illustrate by example and then comment on the advantages of using the CRT.

(1.10)　Example:

As before, consider:

$$7x_1 + 3x_2 = 5$$

$$2x_1 + 3x_2 = 1.$$

This time, use $M = 77 = 7 \cdot 11$. So, $m_1 = 7$, $m_2 = 11$.

1) In Z_7, solving $\bar{A}x^* = b$, we get

$$\begin{pmatrix} 0 & 3 & | & 5 \\ & & | & \\ 2 & 3 & | & 1 \end{pmatrix} \sim \begin{pmatrix} 0 & 3 & | & 5 \\ & & | & \\ 1 & 5 & | & 4 \end{pmatrix}$$

$$\begin{pmatrix} 0 & 1 & | & 4 \\ & & | & \\ 1 & 5 & | & 4 \end{pmatrix} \sim \begin{pmatrix} 0 & 1 & | & 4 \\ & & | & \\ 1 & 0 & | & 5 \end{pmatrix} \sim \begin{pmatrix} 1 & 0 & | & 5 \\ & & | & \\ 0 & 1 & | & 4 \end{pmatrix}$$

So, $x_1^* = \begin{pmatrix} 5 \\ 4 \end{pmatrix}$, $\bar{d}_1 = -2 \cdot 3 = 1$.

2) In Z_{11}:

$$\begin{pmatrix} 7 & 3 & | & 5 \\ & & | & \\ 2 & 3 & | & 1 \end{pmatrix} \sim \begin{pmatrix} 1 & 0 & | & 3 \\ & & | & \\ 0 & 1 & | & 2 \end{pmatrix}$$

$x_2^* = \begin{pmatrix} 3 \\ 2 \end{pmatrix}$, $\bar{d}_2 = 4$.

3) Find d by solving

$$\begin{aligned} d &\equiv 1 \pmod 7 \\ d &\equiv 4 \pmod{11} \end{aligned} \quad \text{with } |d| < \dfrac{77}{2}$$

The solution is $d = 15$.

12

4) Finding $x^* = \begin{pmatrix} \bar{a} \\ \bar{b} \end{pmatrix}$ involves solving

$a \equiv 5 \pmod 7$ $b \equiv 4 \pmod 7$
$a \equiv 3 \pmod{11}$ $b \equiv 2 \pmod{11}$

The CRT gives $a = 47 \pmod{77}$, $b = 46 \pmod{77}$.

Thus, $\quad \bar{y} = \begin{pmatrix} 15 \cdot 47 \\ 15 \cdot 46 \end{pmatrix}$ (coefficients in Z_{77})

$$= \begin{pmatrix} 12 \\ -3 \end{pmatrix} \quad \begin{array}{l} \text{entries reduced to} \\ \text{lie in } \left(\dfrac{-M}{2}, \ \dfrac{M}{2} \right) \end{array}$$

6) So, $y = \begin{pmatrix} 12 \\ -3 \end{pmatrix}$

and $x = \dfrac{1}{d} y = \dfrac{1}{15} \begin{pmatrix} 12 \\ -3 \end{pmatrix}$.

Comments:

1. The advantage of working (mod m_i), for several i, is that the majority of the computations are done in the 'single precision' range determined by the m_i. It is only when using the CRT to pull back to Z_M that multiple precision may be required. The idea is to minimize the use of multiple precision methods which are slower and require greater storage. Use of the CRT method would thus be envisioned when M itself exceeds the computer word-length, yet the m_i are within range.

2. The m_i are usually chosen to be distinct primes, thus trivially satisfying the requirement that $(m_i, m_j) = 1$, for $i \neq j$. Newman [8] suggests that in practice the m_i be chosen as the k largest primes for which the machine allows single precision arithmetic. Note that for the computations for x_i^* to succeed we must have $(d, m_i) = 1$. Newman argues that if the m_i are chosen as large primes the probability that $(d, m_i) > 1$ is small and that the failure of the algorithm due to the singularity of $\bar{A}$ over Z_{m_i} is not an important practical consideration. In any event, were $\bar{A}$ detected to be singular, then the modulus m_i could be dropped and a new one chosen.

Concluding Remarks

For the purposes of exposition our examples have deliberately

13

used small matrices and moduli. The papers [4], [5], [8] contain the results of more typical numerical work. Computer programs that implement the discussed algorithms are contained in [4] and [8]. Lipson [7] gives a discussion of the computing time the algorithms require. These algorithms are useful in handling linear systems whose coefficient matrices are ill-conditioned. In [6] Knuth mentions that for such situations they give "a method for obtaining the true answers in less time than any other known method can produce reliable approximate answers."

References

[1]   E. H. Bareiss, "Computational Solutions of Matrix Problems over an Integral Domain," Journal of the Institute of Mathematics and its Applications, Vol. 10, 1972, p. 68-104.

[2]   I. Borosh and A. S. Fraenkel, "Exact Solutions of Linear Equations with Rational Coefficients by Congruence Techniques," Mathematics of Computation, Vol. 20, no. 93, Jan. 1966, p. 107-112.

[3]   S. Cabay, "Exact Solution of Linear Equations," Proceedings of the 2nd Symposium on Symbolic and Algebraic Manipulation, Los Angeles, 1971.

[4]   J. A. Howell, "Algorithm 406 - Exact Solution of Linear Equations Using Residue Arithmetic," Communications of the ACM, Vol. 14, no. 3, March 1971, p. 180-184.

[5]   J. A. Howell and R. T. Gregory, "An Algorithm for Solving Linear Algebraic Equations Using Residue Arithmetic, I and II," BIT 9 (1969), p. 200-224, p. 324-337.

[6]   D. E. Knuth, The Art of Computer Programming, Vol. 2: Seminumerical Algorithms, Addison-Wesley, Reading, Mass., 1969.

[7]   J. D. Lipson, Elements of Algebra and Algebraic Computing, Addison-Wesley, Reading, Mass., 1981.

[8]   M. Newman, "Solving Equations Exactly," Journal of Research of the National Bureau of Standards, Vol. 71B, no. 4, Oct. - Dec. 1967, p. 171-179.

[9]   N. S. Szabo and R. I. Tanaka, Residue Arithmetic and its Applications to Computer Technology, McGraw-Hill, New York, 1967.

Exercises

1. Solve the simultaneous congruences

    $$x \equiv 2 \pmod{3}$$

    $$x \equiv 4 \pmod{5}$$

    $$x \equiv 2 \pmod{7}$$

    using the method outlined in the constructive version of the CRT.

2. Use the technique of Example (1.9) to solve

    $$3x_1 + x_2 = 1$$

    $$2x_1 + 3x_2 = 2$$

    with 1) M = 17
        2) M = 19.

3. Solve $3x_1 + x_2 = 1$

    $$2x_1 + 3x_2 = 2$$

    using the technique of Example (1.10) with M = 15.

4. Solve $12x_1 + 3x_2 = -1$

    $$-3x_1 - x_2 = -2$$

    with M = 55.

5. Let R be a commutative ring with 1. Assume the following facts about matrices with coefficients in R and their determinants:

    1) $\det(AB) = \det A \ \det B$

    2) $A(adjA) = (adjA)A = (\det A)I$

    Prove:

    1. If A is invertible, then the detA is a unit in R.

    2. If detA is a unit in R, then A is invertible.

6. Let $(a,m) = 1$.

    1) Use the Euclidean algorithm to show that there exist integers x and y with $xa + ym = 1$.

    2) Show that there exists an integer x with $ax \equiv 1 \pmod{m}$.

15

3) Recall Euler's Theorem and examine its connection with 2).

7. Suppose that A is a matrix with coefficients in Z_M (M need not be prime) whose determinant is a unit in Z_M. Think of the entries of A as integers x in the range $0 \leq x < M$.

   1. Define a row-reduction procedure as follows: pick a smallest non-zero element in the first column of A and subtract the row it is in from all other rows containing non-zero elements in the first column. Iterate this procedure until it is no longer possible.

      For example, try it on the matrix
      $$\begin{pmatrix} 3 & 1 \\ 6 & 3 \end{pmatrix} \text{ over } Z_8.$$

   2. Explain why, in general, the end result is a matrix that has exactly one non-zero entry in the first column, that entry being a unit in Z_M.

   3. Explain how this procedure can be used to show that A is row equivalent to the identity matrix. Illustrate it on the matrix $\begin{pmatrix} 2 & 0 & 1 \\ 3 & 1 & 1 \\ 1 & 2 & 1 \end{pmatrix}$ over Z_8.

8. Let the integers a_1, a_2, a_3 be given. Suppose $M = m_1 m_2 m_3$, where $(m_i, m_j) = 1$, if $i \neq j$.

   1) Show that there exist integers c_1, c_2, c_3 so that
      $$x = c_1 + c_2 m_1 + c_3 m_1 m_2$$

      satisfies $x \equiv a_i \pmod{m_i}$, $i = 1,2,3$.

   2) Show that the c_i can be picked with
      $0 \leq c_i < m_i$, $i = 1,2,3$

      Hence, with such c_i, x satisfies $0 \leq x < M$.

      (This gives an alternate constructive proof of the CRT).

2   Block Designs

Block designs arise in statistics in connection with methods for increasing precision and uniformity in the design of experiments. They are tools often used by researchers in situations where the effectiveness of certain treatments is being measured and care must be taken to insure fairness in constructing the experiment. We will focus on methods of constructing block designs using results from elementary group theory and finite field theory.

Example

Researchers are interested in measuring the efficacy of 5 supposed treatments T_1, T_2, T_3, T_4, T_5 for the cure of baldness, but can only administer three of them on a given day. Suppose the following schedule of days and treatments is used.

	DAY	TREATMENTS
	1	T_3 T_4 T_5
	2	T_4 T_5 T_1
	3	T_3 T_1 T_5
	4	T_2 T_4 T_5
(1)	5	T_4 T_2 T_1
	6	T_1 T_2 T_3
	7	T_1 T_3 T_4
	8	T_2 T_3 T_4
	9	T_2 T_3 T_5
	10	T_1 T_2 T_5

This schedule has certain attractive features:
1) The same number (three) of treatments is administered on any day.
2) Each treatment occurs the same number (six) of times.
3) All comparisons of pairs of treatments are of equal importance. More precisely, if we consider any two treatments and count the number of days on which both occur, the same number is always obtained -- three.

Calling the days 'blocks' and the treatments 'objects', the above affords an example of a block design.

Def: A collection of v <u>objects</u> distributed into b sets, called <u>blocks</u>, is a <u>balanced incomplete block design</u> with parameters (v,b,r,k,λ) provided
1) each block contains the same number k of objects
2) each object is in the same number r of blocks

17

        3) each unordered pair of objects occurs in exactly
           λ blocks

Examples and Remarks

1. (1) has   v = 5
              b = 10
              r = 6
              k = 3
              λ = 3.
2. In the interest of brevity, we will refer simply to block designs.
3. 5 objects {0, 1, 2, 3, 4 }in 5 blocks:

(2)
$$\left\{\begin{array}{cccc} 1, & 2, & 3, & 4 \\ 2, & 3, & 4, & 0 \\ 3, & 4, & 0, & 1 \\ 4, & 0, & 1, & 2 \\ 0, & 1, & 2, & 3 \end{array}\right\}$$

is a block design with

              v = 5
              b = 5
              r = 4
              k = 4
              λ = 3.

Note here that the number of objects equals the number of blocks. In
general, if v = b, the block design is called symmetric.

The five parameters of a block design

         v = total number of objects
         b = number of blocks
         r = number of blocks an object appears in
         k = number of objects per block
         λ = number of blocks containing a given pair
               are related.

It is always the case that

(3)   bk = vr

         and

(4)   r(k-1) = λ(v-1).

To establish (3) consider the number of tuples of the form (x, B) where
x is an object and B is a block containing x. Since there are v choices
for x and x appears in r B's, there are thus vr such pairs. On the
other hand, there are b blocks and each block allows k x's. so, there
are bk such pairs. This forces vr = bk.

To establish (4), we again use the fundamental idea of counting some-

                               18

thing two different ways. We will say a 'meeting' occurs between two objects whenever both objects are in a block together. Fix a particular object x and ask, how many meetings is x part of? Since x occurs in r blocks and meets k - 1 objects in each of these blocks, there are r(k - 1) such meetings. From another point of view, x meets each of the other v - 1 objects λ times and hence the number of meetings involving x is λ(v - 1). The equality (4) follows.

For example, a block design with v = 10, b = 12 and k = 5 is impossible.

Constructing block designs with specified parameters by trial and error is a difficult procedure. Our intent is to exhibit several elementary algebraic techniques useful in their construction.

Difference Sets

Consider the symmetric block design of (2) and regard the objects as representing the elements of Z_5. Using the block B = { 1, 2, 3, 4 } there are twelve differences $x_i - x_j$ we can form, where x_i and x_j belong to B, $x_i \neq x_j$. (We are thinking of Z_5 as an additive abelian group). These are

$$\underline{+}(1 - 2) = 1, 4 \qquad\qquad \underline{+}(2 - 3) = 1, 4$$

$$\underline{+}(1 - 3) = 2, 3 \qquad\qquad \underline{+}(2 - 4) = 2, 3$$

$$\underline{+}(1 - 4) = 2, 3 \qquad\qquad \underline{+}(3 - 4) = 1, 4.$$

Note that each of the non-zero elements of Z_5 occurs as a difference of two distinct elements of B exactly three times.
We abstract this.

Def: A set of k different elements D = { x_1, x_2, ..., x_k } from an additive abelian group G of order v is called a (v, k, λ) difference set if and only if for every d ≠ 0, d ε G, there are exactly λ ordered pairs (x_i, x_j), x_i, x_j ε D with $x_i - x_j$ = d.

Examples

1. G = Z_5

    {1, 2, 3, 4 } is a (5, 4, 3) difference set.

2. G = Z_{13}

    {1, 5, 6, 8 } is a (13, 4, 1) difference set.

3. G = Z_6

    {1, 2, 3 } is not a difference set.

Note that in Example (2) each of the five blocks is of the form $\{1 + x, 2 + x, 3 + x, 4 + x\}$ for some x in Z_5. The point is that a difference set always allows this construction.

Proposition 1

Let $G = \{g_0, g_1, \ldots, g_{v-1}\}$ be an additive abelian group of order v. If $D = \{x_1, x_2, \ldots, x_k\}$ is a (v, k, λ) difference set in G, then the sets

$$B_i = \{x_1 + g_i, x_2 + g_i, \ldots, x_k + g_i\}, \quad i = 0, 1, \ldots, v - 1,$$

form a symmetric block design with parameters v = b, k, λ , r = k.

Examples

1. $G = Z_5$

    $B_i = \{1 + i, 2 + i, 3 + i, 4 + i\}$,   i = 0, 1, 2, 3, 4

    gives the block design of Example (2).

2. $G = Z_3$

    $\{1, 2\}$ is a (3, 2, 1) difference set yielding the block design

    $$\begin{matrix} \{1, 2\} \\ \{2, 0\} \\ \{0, 1\} \end{matrix} \quad \text{with} \quad r = 2, \lambda = 1.$$

Better examples will follow.

Proof of the Proposition:

Let $D = \{x_1, x_2 \ldots, x_k\}$ be a (v, k, λ) difference set and x and y be a pair of distinct elements in G. If the pair x, y are to occur together in some block, then we must have

$$x = x_i + g$$

$$y = x_j + g \quad \text{for some i, j and some } g \in G, \text{ and thus}$$

$$x - y = x_i - x_j.$$

Now, there are exactly λ ordered pairs (x_i, x_j) with $x - y = x_i - x_j$. Given any such pair (x_i, x_j), there is a unique $g \in G$ with $x = x_i + g$. Then, $y = x - (x_i - x_j) = (x - x_i) + x_j = x_j + g$. So, the pair x, y occurs in the block indexed by g. Since $g = x - x_i = y - x_j$, distinct pairs (x_i, x_j) yield distinct blocks containing x and y.

20

The designs produced by the above construction have an additional desirable property. We illustrate using again the design of (2).

Say an advertising agency wishes to compare five brands of detergents in tests with subjects. The subjects distinguish the boxes by colors. Along with imposing requirements that are met by using a block design, the agency requires that each brand be given each color the same number of times to eliminate the influence the color of the box may exert. With 5 brands, 4 colors and 5 subjects, the design of (2) satisfies these criteria.

| | | Color | | | |
		Red	Yellow	Green	Blue
	A	1	2	3	4
Subject	B	2	3	4	0
	C	3	4	0	1
	D	4	0	1	2
	E	0	1	2	3

The color restriction is met since in the columns indexed by the colors each object (brand) occurs exactly once. In general, thinking of a column as being indexed by the element x_i in the difference set, all elements of the column are of the form $x_i + x$, $x \in G$, and as x varies these will generate each of the elements of G once. For an actual application of block designs to the testing of dishwashing detergents see [5].

A slight generalization of difference sets allows us to extend Proposition 1.

Examples

1. Consider the two subsets $\{0, 1, 2, 4\}$ and $\{0, 3, 4, 7\}$ of Z_9. Looking at all differences of the form $x_i - x_j \neq 0$ where x_i and x_j are in the same subset, we see

(5)   1) that there are $12 + 12 = 24$ differences and
      2) that among these differences each of the 8 non-zero elements of Z_9 occurs three times.

We will say that the differences arising from the two subsets are symmetrically repeated three times.

2. In Z_{13}, the subsets $\{1, 5, 8, 12\}$, $\{2, 3, 10, 11\}$, and $\{4, 6, 7, 9\}$ yield 36 differences, symmetrically repeated three times.

Suppose we have an abelian group with elements $g_0, \ldots, g_{n-1}$ and a set of t blocks (subsets of G) $B_1, B_2, \ldots, B_t$ such that

21

1) every block has k elements
2) the differences arising from the t blocks are symmetrically
   repeated, each occuring λ times.
then

Proposition 2

The nt blocks

$$B_{j,g_i} = B_j + g_i \quad j = 1, 2, \ldots, t, \quad i = 0, \ldots, n-1$$

give a block design with parameters

$$
\begin{aligned}
v &= n \\
b &= nt \\
r &= kt \\
k & \\
\lambda &
\end{aligned}
$$

Remarks

1. The Proposition applied to (5) above yields a block design with

   $$v = 9, \ b = 18, \ r = 8, \ k = 4, \ \lambda = 3.$$

2. The instance t = 1 of Proposition 2 gives Proposition 1. Since
   their proofs involve similar arguments, the proof of Proposition 2
   is sensibly omitted.

   A reasonable question to ask is, Are there ways to manufacture sets
with symmetrically repeated differences? In the next sections we employ
finite fields and quadratic residues in several such constructions.

Finite Fields

   The arithmetic of the finite fields $GF(p^n)$ can sometimes be ex-
ploited to produce classes of block designs. Here the underlying group
G is the additive group of the field.

Proposition 3

Let $v = 6t + 1 = p^n$, where p is a prime. Let x be a primitive element
of $GF(p^n)$.
Then the blocks

$$\{x^0, \ x^{2t}, \ x^{4t}\}$$

$$\bullet \quad \bullet \quad \bullet$$

$$\{x^i, \ x^{2t+i}, \ x^{4t+i}\}$$

$$\bullet \quad \bullet \quad \bullet$$

$$\{x^{t-1}, \ x^{3t-1}, \ x^{5t-1}\}$$

22

satisfy the conditions of Proposition 2 and produce a block design with

$$v = 6t + 1$$
$$b = t(6t + 1)$$
$$r = 3t$$
$$k = 3$$
$$\lambda = 1$$

Examples:

1. t = 1
    v = 7
    G = GF(7)

We can take x = 3 as a primitive element.

Then

$$\{x^0, x^{2t}, x^{4t}\} = \{1, 2, 4\}$$

with differences

$$\pm(1 - 2) = 1, 6$$

$$\pm(1 - 4) = 3, 4$$

$$\pm(2 - 4) = 2, 5.$$

By Propositions 1 or 2 this produces a symmetric block design with
v = b = 7, r = 3.

2. The t blocks listed in Proposition 3 are called _initial blocks_.

    Let t = 2, v = 13, G = GF(13). Here we take x = 2.
    The initial blocks are

    $$\{2^0, 2^4, 2^8\}, \{2, 2^5, 2^9\}$$

    or {1, 3, 9 }, {2, 6, 5 }.

3. A block design with k = 3 and $\lambda = 1$ is called a _Steiner triple system_.

Proof of the Proposition

For i = 0, ..., t - 1, the six differences in block i are

$$\pm(x^{2t+i} - x^i), \pm(x^{4t+i} - x^i), \pm(x^{4t+i} - x^{2t+i})$$

or

$$\pm x^i(x^{2t} - 1), \pm x^i(x^{4t} - 1), \pm x^i(x^{4t} - x^{2t}).$$

23

Namely, the differences in block zero multiplied by x^i. We show that these 6t differences account for the non-zero elements of $GF(p^n)$.

Now, $x^{6t} = 1.$

So, $x^{6t} - 1 = (x^{3t} - 1)(x^{3t} + 1) = 0.$

Since x is a primitive element,

$$x^{3t} - 1 \neq 0,$$

so that $x^{3t} + 1 = 0.$

i.e. $-1 = x^{3t}$

Also, $x^{2t} - 1 \neq 0.$

So, there exists an s with

$$x^s = x^{2t} - 1.$$

Then $\underline{+}(x^{2t} - 1) = \underline{+}x^s = x^s, x^{s+3t},$

$$\underline{+}(x^{4t} - 1) = \underline{+}(x^{4t} - x^{6t})$$

$$= \underline{+}\, x^{4t}(1 - x^{2t})$$

$$= \underline{+}\, x^{4t}\, x^s = x^{s+4t}, x^{7t}x^s$$

$$= x^{s+t}, x^{s+4t}$$

and $\underline{+}(x^{4t} - x^{2t}) = \underline{+}\, x^{2t}(x^{2t} - 1)$

$$= \underline{+}\, x^{2t}x^s = x^{s+2t}, x^{s+5t}.$$

Therefore, the 6t differences can be written

$$x^s x^i, \; x^s x^{t+i}, \; x^s x^{2t+i}, \; x^s x^{3t+i}, \; x^s x^{4t+i}, \; x^s x^{5t+i},$$

$$i = 0, \ldots, t - 1.$$

But,

$$x^{jt+i}, \; j = 0, 1, \ldots, 5, \quad i = 0, \ldots, t - 1$$

produces all the non-zero elements of the field. Multiplication by x^s simply permutes them. The conditions of Proposition 2 are satisfied with $\lambda = 1$.

24

Example

```
        t =  4
        v = 25
        G = GF(5²)
```

If x is a primitive element, the initial blocks are

$$\{x^0, x^8, x^{16} \}, \quad \{x, x^9, x^{17} \}, \{x^2, x^{10}, x^{18} \},$$

$$\{x^3, x^{11}, x^{19} \}.$$

To produce a primitive element, note that $p(x) = x^2 + 2x + 3$ is an irreducible polynomial over GF(5). Abusing our notation somewhat, we let x be a root of $p(x)$. Then

$$x^2 = -2x + 3 = 3x + 2,$$

$$x^8 = 4x + 1$$

$$x^{12} = 4 = -1.$$

x is thus a primitive element. After some arithmetic, the elements of the initial blocks can be expressed in the form $ax + b$, $a, b \in GF(5)$.

The complete block design can be obtained by adding to the initial blocks the polynomials $ax + b$ (a, b = 0, 1, 2, 3, 4) and reducing coefficients mod 5.

To facilitate notation note that the additive group of GF (5^2) is isomorphic to $Z_5 \times Z_5$.

Writing $ax + b$ as (a,b) or ab (no multiplication intended), the initial blocks can be written

$$\{01, 41, 13 \}, \{10, 33, 12 \}, \{23, 21, 02 \}, \{11, 24, 20 \}$$

and the complete solution can be generated by 'adding mod 5'.

The type of argument used to establish Proposition 3 can be used in other instances. For example,

for $v = 4t + 1 = p^n$, p a prime, the initial blocks

$$\{x^i, x^{t+i}, x^{2t+i}, x^{3t+i} \}, \quad i = 0, 1, \ldots, t - 1,$$

yield a block design with

$$b = t (4t + 1)$$

(6) $$r = 4t$$

$$k = 4$$

$$\lambda = 3.$$

A good exercise awaits the hearty.

25

Quadratic Residues

Some algebraic preliminaries

p will always denote an odd prime.

Let Z_p^* be the multiplicative group of non-zero elements in Z_p.

Now, the mapping $\phi: Z_p^* \longrightarrow Z_p^*$ by $\phi(x) = x^2$ is a homomorphism of abelian groups.

An element $a \in Z_p^*$ is a <u>quadratic residue mod p</u> if $a = \phi(x)$ for some $x \in Z_p^*$.

Alternately, a non-zero integer a with $(a,p) = 1$ is a quadratic residue mod p if there exists an integer x with $x^2 \equiv a \pmod{p}$.

Examples

1) 3 is a quadratic residue (mod 11) since $5^2 \equiv 3 \pmod{11}$

2) The image of ϕ, $\phi(Z_p^*)$, is the <u>set of quadratic residues</u> (mod p). Note that this is a group under multiplication.

3) $\{ 1, 2, 4 \}$ is the set of quadratic residues mod 7.

Proposition 4

The set of quadratic residues mod p has order $\frac{p-1}{2}$.

i. e. exactly half the elements of Z_p^* are quadratic residues.

Proof:

The homomorphism ϕ described above has kernel

$$= \{x \in Z_p^* \mid x^2 = 1\} = \{1, -1\}.$$

Note that $1 \neq -1$ since $p > 2$. Apply the Fundamental theorem of Group Homomorphisms to get

$$|\phi(Z_p^*)| = \frac{|Z_p^*|}{2} = \frac{p-1}{2}.$$

Remarks

1. An element which is not a quadratic residue is called a quadratic non-residue.

2. Proposition 4 shows that $H = \phi(Z_p^*)$ is a normal subgroup of Z_p^* of

26

index two -- so if b is a quadratic non-residue then $Z_p^* = H \cup bH$ gives a coset decomposition of Z_p^* relative to H. Further, if a and b are non-residues, then $aH = bH$ and $abH = (aH)(bH) = (aH)^2 = H$, the last equality courtesy of the fact that the group Z_p^*/H has order two. Thus, $ab \in H$.

Expressed another way,

$\qquad$ (non-residue) $\cdot$ (non-residue) = residue

Our discussion also shows that

$\qquad$ (residue) $\cdot$ (residue) = residue

$\qquad$ (residue) $\cdot$ (non-residue) = non-residue.

Example

Notice that the quadratic residues (mod 7) $\{1, 2, 4\}$ form a difference set since

$$\underline{+}(1 - 2) = 6, 1, \quad \underline{+}(1 - 4) = 4, 3, \quad \underline{+}(2 - 4) = 5, 2.$$

To exhibit a connection between residues and difference sets we need

Lemma

If $p \equiv 3 \pmod 4$, then -1 is not a quadratic residue mod p.

Proof:

Suppose there does exist an integer x with $x^2 \equiv -1 \pmod p$.

Let g be a primitive root (mod p) and write $x \equiv g^t \pmod p$,

Then $-1 \equiv x^2 \equiv g^{2t} \pmod p$

Note that, by hypothesis, $\dfrac{p - 1}{2}$ is odd.

So, $-1 \equiv (-1)^{\frac{p-1}{2}} \equiv g^{(2t)\frac{p-1}{2}} \equiv (g^{p-1})^t \pmod p$.

But, $g^{p-1} \equiv 1 \pmod p$, since g has order p-1.

We then have $-1 \equiv 1 \pmod p$, which is a contradiction.

Our main result is

Proposition 5

Let p be a prime with $p \equiv 3 \pmod 4$.

Write $p = 4t - 1$ for some t.
Let $D = \{a_1, \ldots, a_k\}$ be the set of $k = \frac{p-1}{2}$ quadratic

residues (mod p).

Then D is a (v, k, λ) difference set in Z_p with

$$v = p = 4t - 1$$

$$k = \frac{p-1}{2} = 2t - 1$$

$$\lambda = t - 1.$$

Proof:

Note that -1 is a non-residue (mod p) and hence given any a, b εZ_p^*
exactly one of a - b and b - a is a residue (mod p).
Let $a_i - a_j = x$, where x is a quadratic residue. Then
$x^{-1}a_i - x^{-1}a_j = 1$.
Conversely, if $a_i - a_j = 1$ and x is a quadratic residue then
$(xa_i) - (xa_j) = x$.
This sets up a 1 - 1 correspondence between the set of all pairs
(a_i, a_j) with $a_i - a_j = 1$ and pairs (a_i', a_j') with $a_i' - a_j' = x$ for any
residue x. So any residue occurs as a difference as often as any
other. That each non-residue occurs as a difference $a_i - a_j$ as often as
each residue follows from the fact that

$$a_i - a_j = x \Longleftrightarrow a_j - a_i = - x$$

and that x is a residue if and only if -x is a non-residue. It follows
that D is a difference set with

$$\lambda = \frac{(2t-1)(2t-2)}{p-1}$$

$$= \frac{(2t-1)(2t-2)}{4t-2} = t - 1.$$

Examples

1. The quadratic residues (mod 11) $\{1, 4, 9, 5, 3\}$ yield a symmetric
    block design with v = 11, k = 5 and λ = 2.

2. As a sidelight we note that elementary number theory can be used to
    show that there are infinitely many primes p with $p \equiv 3$ (mod 4).

3. The case t = 2 of Proposition 5 produces the symmetric block design

$$\{1, 2, 4\}$$

{2, 3, 5}

{3, 4, 6}

{4, 5, 0}

{5, 6, 1}

{6, 0, 2}

{0, 1, 3}

for which the picture

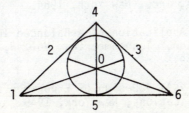

is a nice representation. The blocks are represented by lines (including the circle). In the above any two lines (blocks) intersect in precisely one element. It is a fact that in a symmetric block design any two distinct blocks have exactly λ objects in common (see Exercise 11). Symmetric block designs with λ = 1 thus provide examples of finite projective planes. The example here is often referred to as the Fano geometry.

Concluding Remarks

The subject of block designs has produced an enormous literature. Many of the initial results dealt with statistical analyses of agricultural experiments [6]. Much current research revolves around the question of determining sufficient conditions for the existence of block designs. Equations (3) and (4) are not enough -- there is, for example, no block design with the parameters b = v = 43, k = 7, λ = 1 [3]. Our elementary treatment draws on simplified versions of Bose's "method of symmetrically repeated differences" [1].

Interest in Steiner triples dates back to 1847 when the Rev. Thomas Kirkman posed and solved what is now known as the Kirkman schoolgirl problem: A teacher takes her class of 15 girls on a daily walk. The girls walk in 5 rows of 3 each. The problem is to arrange the schedule so that in 7 consecutive days every girl will have walked in a group of 3 once with every other girl. A solution would require a block design with v = 15, b = 35, r = 7, k = 3, and λ= 1 with the added condition that the design is resolvable into 7 sets of 5 blocks each, with each girl appearing once in each of the 7 sets. For a solution, see [7].

References

[1]   R. C. Bose,   "On the Construction of Balanced Incomplete Block
                      Designs", Annals of Eugenics, 9 (1939), 353-399.

[2]   D. R. Cox,   Planning of Experiments, John Wiley & Sons, Inc.,
                      New York, 1958.

[3]   M. Hall, Jr.   Combinatorial Theory, Blaisdell Publishing Co.,
                      Waltham, 1967.

[4]   M. Hall, Jr.   "Block Designs" in Applied Combinatorial Mathe-
                      matics, Edwin F. Beckenbach editor, John Wiley and
                      Sons, Inc., New York, 1964.

[5]   P. W. M. John, "An Application of a Balanced Incomplete Block
                      Design," Technometrics, Vol. 3, No. 1, 1961, p.
                      51-54.

[6]   H. B. Mann,   Analysis & Design of Experiments, Dover
                      Publications, New York, 1949.

[7]   H. J. Ryser,   Combinatorial Mathematics, Carus Mathematical
                      Monographs No. 14, Mathematical Association of
                      America, 1963.

Exercises

1. Volleyball teams of 5 students each are formed from a group of 15 students. All students play on the same number of teams and any pair of students play together on exactly 2 teams.
   How many teams are there? How many teams is any student on?

2. In a symmetric block design, show that it is always the case that k = r.

3. A <u>tactical configuration</u> is an arrangement of v objects into b blocks so that each of the b blocks contains the same number k of objects and each of the objects lies in exactly r blocks.
   1) Explain why bk = vr.
   2) Give an example of a tactical configuration which is not a block design.

4. Show that $\{0, 1, 5\}$ is a difference set in Z_7. List the parameters of the associated block design.

5. Show that $\{1, 5, 6, 8\}$ is a difference set in Z_{13}. List the parameters of the associated block design.

6. 9 brands of toothpaste are to be tested by 18 subjects, each subject receiving 4 tubes colored in the four colors red, yellow, blue and green. Produce a table showing the distribution of toothpaste in colored tubes among the subjects obeying the conditions that
   1) each brand shall be used by the same number of subjects
   2) each pair of brands shall be used by the same number of subjects
   3) each brand shall be given each color the same number of times.

7. List initial blocks for a block design guaranteed by Proposition 3 in the case t = 3.

8. Explain why in a Steiner triple system v = 6t + 1 or v = 6t + 3 for some t.

9. Fill in the details of Example (6) in the case t = 1.

10. Let g be a primitive element of GF(p). Show that g is a quadratic non-residue (mod p).

11. This exercise is broken into parts meant to guide you through a proof of the interesting fact that in a <u>symmetric</u> block design any two blocks have exactly λ objects in common. For a symmetric block design verify the following.

    1) $v = b$
       $r = k$
       $\lambda(v-1) = r(r-1) = k(k-1)$
       Fix a given block and let $n = \dfrac{r(r-1)}{\lambda} = b - 1$.

    2) The number of pairs formed from this fixed block is

31

$$\frac{r(r-1)}{2} \quad .$$

3) The number of times the above pairs occur in the remaining blocks is $\frac{n \lambda(\lambda - 1)}{2}$.

4) The number of times the objects in the fixed block occur in the other blocks is $n \lambda$.

Let a_1, a_2, ..., a_n be the number of objects common to the fixed block and the remaining n blocks.

5) Then, $a_1 + a_2 + \ldots + a_n = n \lambda$.

6) Also,

$$\frac{a_1(a_1-1)}{2} + \frac{a_2(a_2-1)}{2} + \ldots + \frac{a_n(a_n-1)}{2} = \frac{n \lambda(\lambda-1)}{2}$$

7) So,

$$a_1^2 + a_2^2 + \ldots + a_n^2 = n\lambda^2$$

8) And

$$(a_1 - \lambda)^2 + \ldots + (a_n-\lambda)^2 = 0$$

9) Therefore,

$$a_1 = a_2 = \ldots = a_n = \lambda.$$

12. In a symmetric block design with

   $v = b = 4t - 1$, $k = r = 2t - 1$, and $\lambda = t - 1$,

   given a pair of blocks, how many objects are in neither block?

13. List initial blocks for a design guaranteed by (6) in the case $t = 2$.

3 Error-Correcting Codes I: Hamming Codes

The accurate reception of electronically transmitted data is obviously of major concern. Communications with a satellite marred by a noisy channel can yield faulty information as can a computer relay beset by a component failure. The subject of algebraic coding theory arose as an attempt to use discrete mathematics in the detection and correction of errors incurred in transmitting data. From its roots in the late 1940's and early 1950's it has developed into a major branch of applied mathematics and stands as a prime example of the use of sophisticated abstract algebraic techniques in the solution of practical problems. It affords a very pretty picture of algebra at work.

In this chapter we introduce the notion of coding and error correction via a set of codes first introduced in 1950 by Richard W. Hamming of Bell Telephone Laboratories. The main tools here are elementary linear algebra and the language of cosets. Finite fields and rings will play a significant role in the next chapter that discusses codes with greater error correction capabilities.

In simple terms, we think of a <u>sender</u> who has a list of possible messages to transmit -- this list is the code. At the other end is a <u>receiver</u> who, because of garbled reception, may not receive the exact message that was sent.

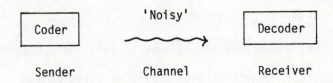

| Coder | 'Noisy' | Decoder |
| Sender | Channel | Receiver |

<u>Our problem:</u>   can the code be designed so that some errors can be detected and, perhaps, corrected?

<u>Example 1:</u>

Suppose there are only two possible messages:

$$0 \ 0 \ 0 \ 0 \quad \text{and} \quad 1 \ 1 \ 1 \ 1$$

are the codewords.

Then if 0 1 0 0 is received, certainly an error has occurred and 'most probably' the original message was 0 0 0 0 -- i.e. we operate under the assumption that it is more probable that a single error has occurred as opposed to three. Later we will clarify our naive use of probability.

If only one error occurrs in transmission the above code is able to detect and correct it -- it is single error correcting. Note though, that receiving 1 1 0 0 would result in ambiguity.

To systematize things we will think of codewords as binary n-tuples. Now, the set of all binary n-tuples forms a vector space over GF(2). Our codes will be subsets of this vector space and, better yet, to introduce some structure, subspaces.

Def.   An <u>(n, k) code</u> is a k-dimensional subspace C of the GF(2) vector space of all binary n-tuples.

<u>Examples and Remarks</u>

1. n is called the <u>length</u> of the code; the elements of the subspace C are referred to as <u>codewords</u>. C is also referred to as a binary linear code.

2. In Example 1, C = $\{(1, 1, 1, 1), (0, 0, 0, 0)\}$ is 1-dimensional subspace of the space of all binary 4-tuples. C is a (4, 1) code.

3. The subspace generated by (1, 1, 0) and (0, 1, 0) in the space of all binary 3-tuples is a (3, 2) code with codewords

$$0 \quad 0 \quad 0$$
$$1 \quad 1 \quad 0$$
$$0 \quad 1 \quad 0$$
$$1 \quad 0 \quad 0$$

4. Note that a (n, k) code has 2^k codewords.

One way of manufacturing subspaces is to look at solution spaces of homogeneous systems of linear equations over GF(2). We think of such systems as being written in the matrix form

(1) $HX = 0.$

If H is a k x n matrix   and

$$X = \begin{pmatrix} x_1 \\ \vdots \\ x_n \end{pmatrix} \quad,$$

this describes a homogeneous system of k linear equations in the unknowns $x_1, \ldots, x_n$.

We recall the fundamental result that the set of all vectors X with $HX = 0$ forms a vector space of dimension $= n - \text{rank } H$, called the solution space of (1).

<u>Example 2:</u>   <u>The (7,4) Hamming code</u>

Consider the 3 x 7 matrix H whose columns are the binary representations of the numbers 1 thorugh $2^3 - 1$,

$$H = \begin{pmatrix} 0 & 0 & 0 & 1 & 1 & 1 & 1 \\ 0 & 1 & 1 & 0 & 0 & 1 & 1 \\ 1 & 0 & 1 & 0 & 1 & 0 & 1 \end{pmatrix}$$

We construct a code C by letting C be the solution space of HX = 0. Now, rank H = 3 (there are clearly three linearly independent columns). So,   dim C = 7 - 3 = 4.

Thus, C is a (7, 4) code with 2^4 = 16 codewords.

Note that X is a codeword if and only if HX = 0. (Here codewords are written as column vectors -- at other times codewords will be listed as row vectors, with or without commas. This is done for notational convenience and should cause no major confusion.)

Thus, 0 1 1 1 1 0 0 is a codeword in C, while 1 0 0 0 0 0 0 is not.

The matrix H above is called the <u>parity-check</u> matrix of the code. In general, H is a parity-check matrix for the code C if C is the solution space of the system HX = 0. We now show that the (7,4) Hamming Code is single error correcting.

Suppose the codeword c is sent and the vector r is received

$$c \xrightarrow{\hspace{3cm}} r$$

and that an error has occurred in one of the components of c, say the ith, changing a 0 to a 1 or vice versa.
Then we can write

$$r = c + e_i$$

where e_i is the binary vector with 0's everywhere except a 1 in the ith position.
<u>e.g.</u>, if

$$c = 0 \ 1 \ 1 \ 1 \ 1 \ 0 \ 0$$

and

$$r = 0 \ 0 \ 1 \ 1 \ 1 \ 0 \ 0,$$

then

$$e_i = 0 \ 1 \ 0 \ 0 \ 0 \ 0 \ 0. \hspace{2cm} (i = 2).$$

Our problem is: knowing r, can we recover c?

Note that knowledge of e_i is sufficient for finding c, since

$$c = r + e_i \hspace{2cm} \text{(remember, all is}$$
$$\text{over GF(2)).}$$

Now,

$$Hr = H(c + e_i)$$
$$= Hc + He_i$$

$$= 0 + He_i \qquad \text{(since c is a codeword)}$$
$$= He_i.$$

So, He_i is known, since r is known.

But,

$$He_i = H \begin{pmatrix} 0 \\ 0 \\ \vdots \\ 1 \\ 0 \\ \vdots \\ 0 \end{pmatrix} \leftarrow \text{ith position}$$

$$= \text{ith column of H.}$$

i.e. knowing He_i, we know the location i of the single error and hence can reconstruct c from r.

Example:

We assume a single error occurs in transmission with the received vector r being r = 0 0 1 1 1 0 0.
The receiver computes

$$Hr = \begin{pmatrix} 0 \\ 1 \\ 0 \end{pmatrix}, \text{ which is the 2nd column of H.}$$

Thus, i = 2 and e_i = 0 1 0 0 0 0 0.
The received vector can then be decoded as the codeword
$$c = r + e_i = 0\ 1\ 1\ 1\ 1\ 0\ 0.$$

Remarks

1. If no error occurs, the receiver recognizes this by noting that $Hr = 0$.

2. The (7, 4) Hamming code will not correct combinations of 2 or more errors:

   If the codeword

   $$0\ 0\ 0\ 1\ 1\ 1\ 1$$

   were received as

   $$r = 0\ 0\ 0\ 0\ 0\ 1\ 1$$

   then

   $$Hr = \begin{pmatrix} 0 \\ 0 \\ 1 \end{pmatrix}.$$

   and $r + e_i = 1\ 0\ 0\ 0\ 0\ 1\ 1$ results in an incorrect decoding. We will shortly see what properties a code must have to be multiple error correcting.

<u>Example 3:</u> <u>The $(2^m - 1, 2^m - 1 - m)$ Hamming Codes</u>

The code of Ex. 2 can be generalized as follows.

For $m \geq 2$, construct the $m \times (2^m - 1)$ matrix H whose ith column is the binary representation of the decimal number i, $1 \leq i \leq 2^m - 1$. The rank of H is m, since H contains m unit columns.

The code C whose parity-check matrix is H is then a $(2^m - 1, 2^m - 1 - m)$ code, also called a Hamming Code. There is thus, <u>e.g.</u>, a (15, 11) Hamming code with $2^{11} = 2048$ codewords. The same argument as used for the (7, 4) code shows that each of these codes is capable of correcting a single error.

The reason H is called a parity-check matrix is because a code-word c must satisfy Hc = 0. This amounts to saying that sums of elements in certain co-ordinates of c must be zero. But a binary sum is determined by whether an odd or an even number of 1's is present.

The role of k in an (n, k) code can also be seen via the parity-check matrix. We use the (7, 4) Hamming code as an example. Write H as

$$H = \begin{pmatrix} 0 & 0 & 0 & 1 & | & 0 & 0 & 0 \\ 0 & 1 & 1 & 0 & | & 0 & 0 & 0 \\ 1 & 0 & 1 & 0 & | & 0 & 0 & 0 \end{pmatrix} \quad + \quad \begin{pmatrix} 0 & 0 & 0 & 0 & | & 1 & 1 & 1 \\ 0 & 0 & 0 & 0 & | & 0 & 1 & 1 \\ 0 & 0 & 0 & 0 & | & 1 & 0 & 1 \end{pmatrix}$$

$$= H_1 + H_2.$$

Then,

$$c = \begin{pmatrix} c_1 \\ c_2 \\ \vdots \\ c_7 \end{pmatrix} \qquad \text{is a codeword}$$

$\Longleftrightarrow$ Hc = 0

$\Longleftrightarrow$ $H_1 c + H_2 c = 0$

$\Longleftrightarrow$ $H_1 c = H_2 c$

$\Longleftrightarrow$ $\begin{pmatrix} 0 & 0 & 0 & 1 \\ 0 & 1 & 1 & 0 \\ 1 & 0 & 1 & 0 \end{pmatrix} \begin{pmatrix} c_1 \\ c_2 \\ c_3 \\ c_4 \end{pmatrix} = \begin{pmatrix} 1 & 1 & 1 \\ 0 & 1 & 1 \\ 1 & 0 & 1 \end{pmatrix} \begin{pmatrix} c_5 \\ c_6 \\ c_7 \end{pmatrix}$

$$\Longleftrightarrow \quad \begin{pmatrix} c_5 \\ c_6 \\ c_7 \end{pmatrix} = \begin{pmatrix} 1 & 1 & 1 \\ 0 & 1 & 1 \\ 1 & 0 & 1 \end{pmatrix}^{-1} \begin{pmatrix} 0 & 0 & 0 & 1 \\ 0 & 1 & 1 & 0 \\ 1 & 0 & 1 & 0 \end{pmatrix} \begin{pmatrix} c_1 \\ c_2 \\ c_3 \\ c_4 \end{pmatrix}$$

$$\Longleftrightarrow \quad \begin{pmatrix} c_5 \\ c_6 \\ c_7 \end{pmatrix} = \begin{pmatrix} c_2 + c_3 + c_4 \\ c_1 + c_3 + c_4 \\ c_1 + c_2 + c_4 \end{pmatrix}$$

Thus a codeword is obtained by freely choosing the first four digits (called the information digits) c_1, c_2, c_3, c_4 and then picking c_5, c_6, c_7 (called the check digits) according to the relations above. We could view the coding process as starting with a random 4 digit message (vector) which is then encoded into a 7 digit vector for transmission purposes. Intuitively, the check digits are the redundancy necessary to allow error correction. A similar argument can be given to show that any (n, k) code vector can be separated into k information digits and n - k check digits, but since subleties involving permutations of coordinates crop up in a general argument we content ourselves with this example.

Hamming Distance and Error Correction

The error correction procedure used above is an example of a technique called maximum likelihood decoding which rests on the notion of decoding a received vector as that codeword which is in some sense closest to it. We make this more precise by introducing a distance between binary n vectors.

Def. The weight, $w(x)$, of a binary n-tuple x is the number of non-zero components in x. The (Hamming) distance $d(x, y)$ between two binary n-tuples is the number of components they differ in.

Example:

$$w(1\ 1\ 0) = 2$$
$$d(1\ 1\ 0,\ 1\ 0\ 1) = 2$$
$$w(1\ 0\ 1) = d(1\ 0\ 1,\ 0\ 0\ 0) = 2$$
$$d(\ 1\ 1,\ 1\ 1) = 0$$

For binary n-tuples x and y it is easy to see that

(2) $\qquad d(x,\ y) = w(x + y).$

Example:

$$d(0\ 0\ 1\ 1,\ 0\ 1\ 1\ 0) = w(0\ 0\ 1\ 1 + 0\ 1\ 1\ 0)$$
$$= w(0\ 1\ 0\ 1) = 2.$$

An (n, k) code has finitely many codewords, so that it makes sense to speak of the closest that two distinct codewords could be.

Def: The minimum distance d^* of a code C is given by

38

$$d* = \min \{d(x,y) \mid x, y \ \varepsilon \ C, \ x \neq y\}.$$

Example:

For the code of Ex. 1, d* = 4.

By (2), any distance is a weight. Also, w(x) = d(x, 0) says that any weight is a distance. Thus, d* can alternately be described as the smallest weight achieved by a non-zero codeword.

That is,

Proposition 1

$$d* = \min\{w(x) \mid x \ \varepsilon \ C, \ x \neq 0\}.$$

What is d* for the Hamming Codes? The following will be useful.

Proposition 2

Let C be a code with parity-check matrix H. Then d* > ℓ if and only if every set of ℓ columns of H is linearly independent.

Proof:

Suppose that every set of ℓ columns of H is linearly independent but that d* $\leq \ell$. Then by Proposition 1 there is a codeword c ε C with $0 < w(c) = k \leq \ell$. So, c has 1's in k positions, say i_1, ..., i_k. Let H_{i_1}, H_{i_2}, ..., H_{i_k} denote the corresponding columns of H. Since c is a codeword, Hc = 0. Written out this says

$$H_{i_1} + H_{i_2} + \ldots + H_{i_k} = 0.$$

Thus the set of k columns H_{i_1}, ..., H_{i_k} is linearly dependent.

If need be, randomly add other columns to create a set of ℓ linearly dependent columns of H. This violates the hypothesis and produces a contradiction. The proof of the converse is left as an exercise.

Example:

Since any two columns of the parity-check matrix of the (7, 4) Hamming code are distinct and hence linearly independent, we must have d* > 2. Since 1 1 1 0 0 0 0 is a codeword of weight 3, this forces d* = 3.

A similar argument shows that the minimum distance between non-zero codewords in any $(2^m - 1, 2^m - 1 - m)$ Hamming code is 3.

We now establish the connection between the minimum distance of a code C and its error correcting capabilities.

Def.   Let x ε C, r $\geq$ 0. The <u>sphere of radius r</u> about x, S(x, r), is given by

$$S(x, r) = \{y \mid y \text{ is a binary n-tuple with } d(x,y) \leq r \}$$

i.e. S(x, r) consists of those n-tuples that differ from x in at most r positions.

Example:

S(1 0 0, 1) = {1 0 0, 0 0 0, 1 1 0, 1 0 1 }

It is useful to draw on geometric intuition and to think of spheres about codewords pictorially. For example, if c and c' are codewords then the two spheres of radius r about each of them are disjoint if and only if d(c, c') > 2r

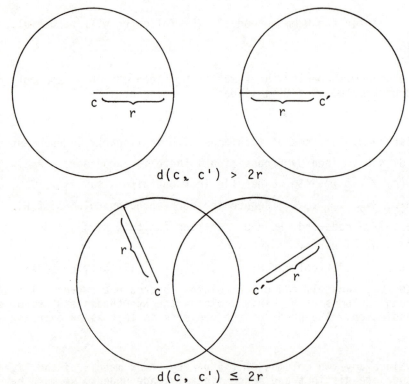

d(c, c') > 2r

d(c, c') ≤ 2r

We will use the following decoding rule, called maximum likelihood decoding:

a received word is decoded as that codeword which is closest to the received word in Hamming distance. Ties are settled arbitrarily.

The question of why this is a reasonable procedure will be addressed shortly, though the very terminology suggests an answer.

Def: A code is t-error correcting if application of maximum likelihood decoding results in correct decoding of any received word which differs from the sent codeword in t or fewer places.

40

Proposition 3

A code C is t-error correcting if and only if its minimum distance is greater than 2t (i.e. if $d^* \geq 2t + 1$).

Proof:

Saying a code is t-error error-correcting is the same as saying that spheres of radius t about distinct codewords are disjoint.

For example, if overlap did occur,

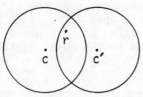

r would be decoded as c, though it may have originated from c'.

Our remarks before say this requirement is equivalent to $d(c, c') > 2t$, for all distinct codewords c and c'; namely $d^* > 2t$.

Remarks

1. Since $d^* = 3$ for the Hamming codes, we see again that they are capable of single error correction. In the event a single error has occurred maximum likelihood decoding is the same as the error correction procedure introduced earlier for the Hamming codes.

2. Though we have one for the Hamming codes, in general the proposition provides no effective procedure for carrying out the decoding short of taking a received vector and computing the distance between it and all possible codewords to find the closest one. For an (n, k) code this requires 2^k computations, which can be prohibitive for large values of k. A significant portion of coding theory is concerned with finding decoding procudures that can be reasonably implemented.

3. In a t-error correcting code, spheres of radii t about codewords are disjoint. Given a codeword c of length n, there are $\binom{n}{k}$ binary n-tuples that differ from c in exactly k positions. Thus, the sphere of radius t about c contains

$$\binom{n}{0} + \binom{n}{1} + \binom{n}{2} + \ldots + \binom{n}{t} \text{ n-tuples.}$$

    If M is the number of codewords in the code then, noting that there are 2^n binary n-tuples in all, we must have

    (3) $M \left[\binom{n}{0} + \binom{n}{1} + \binom{n}{2} + \ldots + \binom{n}{t} \right] \leq 2^n$,

    since the number of vectors accounted for by the spheres cannot exceed the total number of vectors in the space. (3) is sometimes called the Hamming bound, since it gives an upper limit on the number of codewords M, in a t-error correcting code of length n.

    A code is called perfect if equality holds in (3). Equivalently, a t-error correcting code is perfect if every n-tuple is in some sphere of radius t about a codeword.

41

A (7, 4) Hamming code is perfect since

$$2^4 \left[\binom{7}{0} + \binom{7}{1} \right] = 2^7$$

($n = 7$, $t = 1$, $M = 2^4$).

Perfect codes are relatively rare. It was shown (1973) that the Hamming codes and a (23, 12) triple error correcting code discovered by the engineer Golay (1949) are the only possible perfect linear binary codes.

4. Propositions 2 and 3 combined say that the parity-check matrix of a t-error correcting code must have all sets of 2t of its columns linearly independent. Examples of multiple error correcting codes (i.e., t > 1) will be given in the next chapter.

The Rationale for Maximum Likelihood Decoding

We adopt a model called the binary symmetric channel (BSC) to discuss probabilities connected with decoding. In transmitting a string of 0's and 1's we assume that there is a probability p that a given symbol will be received in error, So, e.g., if 0 is sent, 0 will be received with probability 1 - p. This is summarized in the diagram

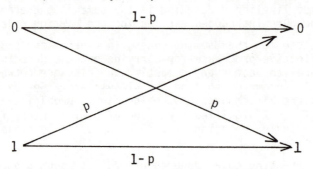

Binary Symmetric Channel (BSC)

A choice of p = .001, e.g., would amount to the observation that our channel corrupts roughly one in a thousand symbols. We always assume that $p < \frac{1}{2}$. In actual practice relatively small values of p occur. We also make the assumption that errors occur randomly and independently -- that the channel has no memory.

Say the codeword c is sent while r is received. The decoder asks, which codeword was most likely to have been sent, given that r was received.

Define the error vector e by r = c + e. Then e has a 1 precisely in those positions where errors have occurred, and 0's elsewhere. Knowledge of e suffices for the decoder, since

c = r + e.

The decoder then aims to choose the most likely e, given r.

In the BSC an error occurs with probability p. So, for example, in a code of length 4

$$prob(e = 0\ 0\ 0\ 0) = \text{probability that no errors}$$
$$\text{have occurred} = (1 - p)^4$$

$$prob(e = 1\ 0\ 0\ 0) = p(1 - p)^3$$

$$prob(e = 1\ 0\ 0\ 1) = p^2 (1 - p)^2.$$

In general, in a code of length n, the probability that e is a given fixed vector of weight ℓ is

$$p^\ell (1 - p)^{n-\ell} .$$

Since $p < \frac{1}{2}$, we have $1 - p > p$. So,

$$(1 - p)^n > p(1 - p)^{n-1} > p^2(1 -1)^{n-2} > \ .\ .\ .\ > p^n.$$

Thus, no errors are more probable than any given error vector of weight 1, which in turn is more probable than any of weight 2, and so on. Under the assumption that all codewords are equally likely, the decoder's best strategy then is to choose an error vector e of smallest weight such that r + e is a codeword. That is, decode r as the codeword nearest to r in Hamming distance. This then maximizes the likelihood of successful decoding.

Decoding Using Cosets

A code C is a subspace (additive subgroup) of the space V_n of all binary n-tuples. Another view of maximum likelihood decoding is gotten by considering the cosets of C in V_n.

A typical coset of C is of the form C + x, where x εV_n. If dim C = k, then there are 2^{n-k} distinct cosets of C in V_n.

If the received vector r is written as

$$r = c + e,$$

where c is the sent codeword, then

$$e = c + r.$$

That is, e is an element of the coset C + r determined by r.

Maximum likelihood decoding then reduces to finding a vector e of least weight in the coset determined by r.

A vector of least weight in a coset is called a coset leader.

The decoding procedure can then be summarized as:

1)   find the coset r belongs to

2)   find a coset leader e in that coset

3)   decode r as r + e.

In the presence of a parity-check matrix H, 1) of the procedure can be accomplished by observing

Proposition 4

The binary n-tuples x and y are in the same coset of C if and only if

Hx = Hy.

Proof:

x and y are in the same coset

$\Longleftrightarrow$ x - y is an element of C

$\Longleftrightarrow$ H(x - y) = 0

$\Longleftrightarrow$ Hx = Hy.

For a binary n-tuple x, Hx is called the _syndrome_ of x. So, distinct syndromes correspond to distinct cosets of C and knowing the syndrome of a received vector r identifies the coset r belongs to.

A way of implementing the above decoding procedure would be to make a list of possible syndromes (one for each coset) and place next to each syndrome a coset leader corresponding to the syndrome. Such a list is called a _decoding table_.

Example 5:

Consider the (5, 2) code

$$C = \{0\,0\,0\,0\,0,\ 1\,1\,1\,0\,0,\ 0\,1\,1\,1\,1,\ 1\,0\,0\,1\,1\}$$

$$H = \begin{pmatrix} 1 & 1 & 0 & 1 & 0 \\ 0 & 1 & 1 & 0 & 0 \\ 0 & 0 & 0 & 1 & 1 \end{pmatrix} \text{ is a parity-check matrix for C.}$$

There are $2^{5-2} = 2^3$ cosets of C in V_5, each containing 4 vectors. A list of these cosets can be gotten by writing all the elements of C in one row -- this is the coset C = C + 0. Then pick a vector x_1 in V_5 not in the first row, and let the second row be the coset C + x_1. Continue this by picking a vector x_2 not in either of the first two rows. Then let C + x_2 be the third row, and so on. If at each stage, the vector x_i is chosen to be a vector of _least_ weight which is not present in the first i rows, then x_i is guaranteed to be a coset leader. One possible such listing for the code C is

				Coset
0 0 0 0 0	1 1 1 0 0	0 1 1 1 1	1 0 0 1 1	C
1 0 0 0 0	0 1 1 0 0	1 1 1 1 1	0 0 0 1 1	C + 1 0 0 0 0
0 1 0 0 0	1 0 1 0 0	0 0 1 1 1	1 1 0 1 1	C + 0 1 0 0 0
0 0 1 0 0	1 1 0 0 0	0 1 0 1 1	1 0 1 1 1	C + 0 0 1 0 0

```
0 0 0 1 0     1 1 1 1 0     0 1 1 0 1     1 0 0 0 1     C + 0 0 0 1 0

0 0 0 0 1     1 1 1 0 1     0 1 1 1 0     1 0 0 1 0     C + 0 0 0 0 1

0 0 1 1 0     1 1 0 1 0     0 1 0 0 1     1 0 1 0 1     C + 0 0 1 1 0

0 1 0 1 0     1 0 1 1 0     0 0 1 0 1     1 1 0 0 1     C + 0 1 0 1 0
```

A <u>decoding table</u> for C would then be

Syndrome	Coset Leader
0 0 0	0 0 0 0 0
1 0 0	1 0 0 0 0
1 1 0	0 1 0 0 0
0 1 0	0 0 1 0 0
1 0 1	0 0 0 1 0
0 0 1	0 0 0 0 1
1 1 1	0 0 1 1 0
0 1 1	0 1 0 1 0

Thus, if r = 1 1 1 0 1 were received, the decoder would compute the syndrome Hr = 0 0 1. This corresponds to the coset leader 0 0 0 0 1 and hence r would be decoded as the codeword r + e = 1 1 1 0 0.

For an (n, k) code a decoding table requires a list of syndromes and coset leaders corresponding to the 2^{n-k} cosets of C. This may or may not be less expensive than checking a received vector against all possible 2^k codewords and picking the nearest one. In a (100, 80) code, for example, there are only 2^{20} cosets as opposed to 2^{80} codewords -- though 2^{20} is still a very big number. A decoding table for a (72, 64) code would involve $2^8 = 256$ cosets, a manageable number. In Example 5 comparing a received vector with the 4 codewords would be simpler than using a decoding table -- the example was contrived as an illustration of coset decoding.

Coset decoding is really an expression of the fundamental theorem of vector space (or group) homomorphisms. If T: $V \longrightarrow W$ is a linear transformation between the vector spaces V and W, the theorem states that the image of T is isomorphic to the quotient space $V/_K$, where K is the kernel (null space) of T:

$$V/_K \simeq \text{Image of T}.$$

The required translation is as follows:
Take
$$V = V_n$$
and

45

$$T(x) = Hx,$$

where H is a parity-check matrix.

Then the kernel K is the code C, while the image of T is the set of syndromes. The fundamental theorem then says that the syndromes are in 1 - 1 correspondence with the cosets of K = C.

Concluding Remarks

References for coding theory are contained at the end of the next chapter. We have confined our attention to binary codes -- codes over GF(2). It is easy to generalize the notion of an (n, k) code to an arbitrary finite field GF(q). For example, one of the few known perfect codes is an (11, 6) double error correcting code over GF(3).

In our discussion of the BSC we have assumed that errors occur independently of each other. Since in many concrete situations this is not so, attention has also been given to codes that deal with channels where errors occur in bursts or clusters.

We should note that error-correcting codes have little in common with cryptographic codes. One set of codes choses to reveal, the other to conceal. The bibliography in the text by McWilliams and Sloane lists many articles that reveal the range of applications of coding theory. Hamming's original paper appeared in the Bell System Technical Journal 29, (1950), p. 147-160 and is very readable.

Exercises

1. With the (7, 4) Hamming code, assuming that at most one error has occurred, decode:

    a) r = 1 1 0 0 0 0 0

    b) r = 1 1 1 1 1 1 1

    c) r = 0 1 0 0 0 0 0

2. Let C be the (15, 11) Hamming code.

    a) Give an example of two non-zero codewords in C.

    b) Can C correct all combinations of two or fewer errors?

3. Show that $d^* = 3$ for the $(2^m - 1, 2^m - 1 - m)$ Hamming code.

4. Show that the $(2^m - 1, 2^m - 1 - m)$ Hamming codes are perfect.

5. Prove the converse of Proposition 2.

6. A generator matrix for an (n, k) code C is a k x n matrix whose rows are a basis for the subspace C.

    a) Find a generator matrix for the (7, 4) Hamming code.

    b) Suppose a code has generator matrix
        $$\begin{pmatrix} 1 & 1 & 0 & 1 \\ 0 & 1 & 1 & 1 \end{pmatrix}.$$

Use maximum likelihood decoding to decode r = 0 1 1 0. Is this code single error-correcting?

7.   List a decoding table for the (7, 4) Hamming code.

8.   Verify the calculations in Example 5. Is the code of Example 5 single/double error-correcting? Is it perfect?

9.   Consider the (5, 3) code with parity-check matrix
$$H = \begin{pmatrix} 1 & 1 & 1 & 1 & 1 \\ 0 & 0 & 0 & 1 & 1 \end{pmatrix}.$$
     a) List the cosets of C and find a decoding table.

     b) Decode the received vector 1 0 0 1 0.

     c) Is this code single error correcting?

10. Construct a (5, 2) single error-correcting code.

11. Can a (6, 3) code be double error-correcting? Why?

12. Binary codes can be generalized to codes that are vector spaces over any finite field GF(q). We show how Hamming codes can be constructed over the field GF(3).

     a) Let H be a matrix with entries from GF(3) whose columns are all possible non-zero m-tuples over GF(3) such that no two columns are scalar multiples of one another.

         Write down such an H in the case m = 2. In general, show that H has $\dfrac{3^m - 1}{2}$ columns

     b) Let C be the code with parity-check matrix H. (A codeword is an n-tuple of elements from GF(3)).

     c) What is the minimum distance d* for this code? (Hamming distance is defined in the same way.)

     d) How many elements are in the sphere of radius 1 about a codeword?

     e) Show that C is a perfect code

The curious reader might want to repeat this exercise with GF(3)

replaced by GF(q).

4 Error-Correcting Codes II: BCH Codes

The previous chapter introduced basic notions of coding along with the single error-correcting Hamming codes. Continuing the development we now show how the machinery of finite fields can be used to construct codes that are multiple error-correcting. To motivate later ideas we take another look at the Hamming codes.

Introduction: An alternate description of the (7, 4) Hamming code.

The columns of the parity check matrix H of the (7, 4) Hamming code are binary vectors of length 3. As such, they can be thought of as polynomials of degree ≤ 2 over GF(2).

For example,

$$\begin{pmatrix} 1 \\ 0 \\ 1 \end{pmatrix} \qquad \text{corresponds to} \quad x^2 + 1.$$

while

$$\begin{pmatrix} 0 \\ 1 \\ 0 \end{pmatrix} \qquad \text{corresponds to} \quad x.$$

This suggests that the columns of H can be indexed by elements of the field GF(2^3). More precisely, let α be a primitive element of GF(2^3) -- say, take α to be a root of the GF(2) polynomial $x^3 + x + 1$.
Then the 7 non-zero elements of GF(2^3) are the powers α^i, i = 0, . . . , 6.
Alternately, elements of GF(2^3) can be written as polynomials in α of degree ≤ 2 and can thus be thought of as binary columns of length 3.

This correspondence for elements of GF(2^3) is made explicit in the following table.

Table 1

α^0	1	0 0 1
α	α	0 1 0
α^2	α^2	1 0 0
α^3	$1 + \alpha$	0 1 1
α^4	$\alpha + \alpha^2$	1 1 0
α^5	$1 + \alpha + \alpha^2$	1 1 1
α^6	$1 + \alpha^2$	1 0 1

Using this corresondence, by $(1 \; \alpha \; \alpha^2 \; \alpha^3 \; \alpha^4 \; \alpha^5 \; \alpha^6)$ is meant the matrix whose i^{th} column is α^{i-1} represented as a binary vector.

So

$$(1 \ \alpha \ \ldots \ \alpha^6) \ = \ \begin{pmatrix} 0 & 0 & 1 & 0 & 1 & 1 & 1 \\ 0 & 1 & 0 & 1 & 1 & 1 & 0 \\ 1 & 0 & 0 & 1 & 0 & 1 & 1 \end{pmatrix}$$

and this, but for the order of the columns, is the parity check matrix for the (7, 4) Hamming code. We worry not about this rearrangement of columns, since the only effect on the code is to correspondingly permute letters of codewords.

Likewise, codewords in the (7, 4) Hamming code can be written as binary polymonials of degree ≤ 6, where the codeword $(a_0, \ldots, a_6)$ corresponds to the polynomial

$$a_0 + a_1 x + a_2 x^2 + \ldots + a_6 x^6.$$

Now,

$$c \ = \ (a_0, \ldots, a_6)$$

is a codeword

$$\longleftrightarrow \quad H \begin{pmatrix} a_0 \\ \vdots \\ a_6 \end{pmatrix} = 0 \ \longleftrightarrow \ (1 \ \alpha \ \ldots \alpha^6) \begin{pmatrix} a_0 \\ \vdots \\ a_6 \end{pmatrix} = 0$$

$$\longleftrightarrow \quad a_0 + a_1 \alpha + \ldots + a_6 \alpha^6 = 0$$

$$\longleftrightarrow \quad f(\alpha) \ = \ 0, \ \text{where} \ f(x) \ = \ a_0 + \ldots + a_6 x^6.$$

So, thinking of codewords as binary polymonials, the (7, 4) Hamming code consists of all polymonials $f(x)$ of degree ≤ 6 having α as a root.

Now, $m(x) \ = \ x^3 + x + 1$ is the minimum polymonial for α over $GF(2) \ = \ Z_2$, and <u>all</u> polynomials having α as a root are precisely the elements of the ideal $<m(x)>$ generated by $m(x)$ in the polynomial ring $Z_2[x]$. Since α is an element of order 7, α is a root of $x^7 - 1$ and hence $m(x)$ divides $x^7 - 1$. Consider the quotient ring

$$Z_2[x] \big/ {}_{<x^7 - 1>}$$

and the image of the ideal <m(x)> in this quotient. Given any
multiple $\ell(x)m(x)$ of $m(x)$, we divide by $x^7 - 1$ and write
$\ell(x)m(x) = q(x)(x^7 - 1) + r(x)$, where $r(x)$ is of degree ≤ 6.
$\ell(x)m(x)$ and $r(x)$ determine the same coset in the quotient ring. Also,
$r(x) = \ell(x)m(x) - q(x)(x^7 - 1)$, and since $m(x)$ divides $(x^7 - 1)$, it
follows that $r(x)$ is a multiple of $m(x)$. This says that the image of
the ideal <m(x)> under the quotient map from $Z_2[x]$ to $Z_2[x]/_{<x^7 - 1>}$
can be viewed as those polynomials $f(x)$ divisible by $m(x)$ of degree
≤ 6. We thus arrive at a description of the (7, 4) Hamming code as the
ideal generated by the image of $m(x)$ in the quotient $Z_2[x]/_{<x^7 - 1>}$.

If the above argument makes sense to you, you should should be
willing to accept that if α is a primitive element of $GF(2^m)$, $m(x)$ is
the minimum polynomial of α over Z_2, and $n = 2^m - 1 =$ the multiplicative order of α, then the $(2^m - 1, 2^m - 1 - m)$ Hamming code can
be viewed as the ideal generated by the image of $m(x)$ in
$Z_2[x]/_{<x^n - 1>}$.

Cyclic Codes

The reason for providing an abstract view of Hamming codes was
to suggest a general way of constructing codes.

Let F be a finite field, n a positive integer. Then the
quotient ring

$$F[x]/_{<x^n - 1>}$$

is an n-dimensional vector space over F. The ideals I of this
quotient ring are vector subspaces and are in 1 - 1 correspondence with
polynomials $g(x)$ that divide $x^n - 1$ in $F[x]$. Such an ideal I is the
image of the ideal <g(x)> in F [x] under the quotient map

$$F[x] \rightarrow F[x]/_{<x^n - 1>}.$$

The individual elements of I are in 1 - 1 correspondence with
polynomials of degree $\leq n - 1$ that are multiples of $g(x)$, and such
elements can be thought of as n-tuples over F.

We think of such an ideal as a code C. That is, given a

51

polynomial g(x) that divides $x^n - 1$ in $F[x]$,

$$C = \{a(x)g(x) | \deg a(x)g(x) \leq n - 1, a(x) \in F[x] \}.$$

A code constructed in this way is called <u>cyclic</u> (See Exercise 7 for the reason behind this terminology).

The polynomial g(x) is called a <u>generator polynomial</u> for the code. If g(x) has degree k, then

$$\dim C = n - k$$

since

$$g(x), xg(x), \ldots, x^{n-k-1}g(x)$$

form a basis for the code.

Example 1

$$F = GF(2), n = 15, g(x) = x^3 - 1.$$

Note that $x^3 - 1$ divides $x^{15} - 1$ over GF(2). The generator polynomial g(x) determines a (15, 12) code.

At times it is more convenient to specify the generator polynomial indirectly. Suppose $\alpha_1, \ldots, \alpha_\ell$ are distinct elements in some extension of a base field F. Then the set of all polynomials $f(x) \in F[x]$ with

$$f(\alpha_i) = 0, \quad i = 1, \ldots, \ell$$

-- <u>i.e.</u> that have all the α's as roots -- forms an ideal in $F[x]$. Let $m_i(x)$ be the minimum polynomial of α_i over F. Then $m_i(x)$ must divide any such f(x), $i = 1, \ldots, \ell$. Hence, if f(x) is in the ideal, then f(x) must be divisible by the least common multiple of the $m_i(x)$'s. That is, the ideal is generated by

$$g(x) = \text{lcm}(m_1(x), \ldots, m_\ell(x)).$$

To allow the previous construction, we need an n for which g(x) divides $x^n - 1$. We claim

$$n = \text{lcm of the orders of the } \alpha_i$$

will do the job. Since then $\alpha_i^n = 1$ and α_i satisfies $x^n - 1$. So

52

$m_i(x)$ divides $x^n - 1$ for all i, and hence $x^n - 1$ is divisible by the least common multiple, $g(x)$.

Choosing n in this way, $g(x)$ is then the generator polynomial of an n - deg $g(x)$ dimensional code over F.

Example 2

$F = GF(2)$. Let α be a root of $x^4 + x + 1$ over $GF(2)$. α is a primitive element of $GF(2^4)$.

Let $\qquad \alpha_1 = \alpha, \alpha_2 = \alpha^3, \alpha_3 = \alpha^5.$

We specify a code by considering all polynomials $f(x)$ of degree $\leq n - 1$ with

$$f(\alpha) = f(\alpha^3) = f(\alpha^5) = 0.$$

Here, $n = \text{lcm}(15, 5, 3) = 15$.

Then, $m_1(x) = x^4 + x + 1$, since this polynomial is irreducible over $GF(2)$.

Since α^3 is a root of $m_2(x)$, $(\alpha^3)^2$, $(\alpha^3)^4$ and $(\alpha^3)^8$ must also be roots of $m_2(x)$ (due to the fact that the mapping $x \longrightarrow x^2$ is an automorphism of $GF(2^4)$ fixing $GF(2)$).
I.e.,

$$\alpha^3, \alpha^6, \alpha^{12}, \alpha^{24} = \alpha^9$$

must be roots.

But $m_2(x)$ can have degree at most 4. This forces
$$m_2(x) = (x - \alpha^3)\,(x - \alpha^6)\,(x - \alpha^{12})\,(x - \alpha^9).$$

Doing some arithmetic with the aid of Table 2, we get
$$m_2(x) = x^4 + x^3 + x^2 + x + 1.$$

Likewise, the roots of $m_3(x)$ must include α^5, $(\alpha^5)^2$, $(\alpha^5)^4$, $(\alpha^5)^8$.
Only two of these are distinct: α^5, α^{10}.

Note that

53

$$(x - \alpha^5)(x - \alpha^{10}) = x^2 + x + 1$$

which is irreducible over GF(2).

Thus,

$$m_3(x) = x^2 + x + 1.$$

Alternately, we could observe that α^5 is an element of order 3 and must thus satisfy

$$x^3 - 1 = (x - 1)(x^2 + x + 1)$$

and the second factor yields the minimum polynomial of α^5.

So,

$$g(x) = \text{lcm}(m_1(x), m_2(x), m_3(x))$$

$$= (x^4 + x + 1)(x^4 + x^3 + x^2 + x + 1)(x^2 + x + I)$$

(since the $m_i(x)$ are
pairwise relatively
prime)

$$= x^{10} + x^8 + x^5 + x^4 + x^2 + x + 1$$

Thus $g(x)$ is the generator polynomial of a $(n, n-\deg g(x)) = (15, 5)$ code over GF(2).

The codeword $xg(x)$ would be represented by the binary string

$$0\ 1\ 1\ 1\ 0\ 1\ 1\ 0\ 0\ 1\ 0\ 1\ 0\ 0\ 0.$$

54

Table 2

The field GF(2^4)

α is a root of $x^4 + x + 1$ over GF(2) and is a primitive element.

$\alpha^0 = 1$ $\qquad\qquad$ $\alpha^{10} = \alpha^2 + \alpha + 1$

$\alpha = \alpha$ $\qquad\qquad$ $\alpha^{11} = \alpha^3 + \alpha^2 + \alpha$

$\alpha^2 = \alpha^2$ $\qquad\qquad$ $\alpha^{12} = \alpha^3 + \alpha^2 + \alpha + 1$

$\alpha^3 = \alpha^3$ $\qquad\qquad$ $\alpha^{13} = \alpha^3 + \alpha^2 + 1$

$\alpha^4 = \alpha + 1$ $\qquad\qquad$ $\alpha^{14} = \alpha^3 + 1$

$\alpha^5 = \alpha^2 + \alpha$ $\qquad\qquad$ $\alpha^{15} = 1$

$\alpha^6 = \alpha^3 + \alpha^2$

$\alpha^7 = \alpha^3 + \alpha + 1$

$\alpha^8 = \alpha^2 + 1$

$\alpha^9 = \alpha^3 + \alpha$

Example 3

Let α be a primitive element of $GF(2^6)$ and $\beta = \alpha^3$. Then β has order 21. Consider the code determined by all polynomials $f(x)$ of degree $\leq n - 1$ that have β and β^3 as roots.

Here

$$n = \ell cm(21, 7) = 21.$$

The minimum polynomial $m_1(x)$ of β has precisely the roots

$$\beta, \beta^2, \beta^4, \beta^8, \beta^{16}, \beta^{32} = \beta^{11}$$

and hence is of degree 6.

The roots of the minimum polynomial $m_2(x)$ of β^3 are

$$\beta^3, \beta^6, \beta^{12};$$

so $m_2(x)$ has degree 3.

Thus the generator polynomial

$$g(x) = m_1(x)m_2(x)$$

determines a (21, 12) binary code.

BCH Codes

A general class of cyclic codes capable of multiple error correction was discovered independently by Bose and Chaudhuri (1960) and Hocquenghem (1959). These are now referred to as BCH codes. They are some of the most oft used codes in channels where errors occur independently. Their importance was enhanced by discovery of efficient decoding procedures that allowed for their practical implementation.

56

The general definition goes this way:

Let m_0, d be integers, $\alpha \in GF(q^m)$.

The cyclic code consisting of all vectors $f(x)$ of degree $\leq n - 1$ over GF(q) for which

$$\alpha^{m_0}, \alpha^{m_0+1}, \ldots, \alpha^{m_0+d-2}$$

are roots of $f(x)$ is a BCH code.

The <u>most important BCH codes</u> arise when

α is a primitive element of $GF(2^n)$, $m_0 = 1$ and $d = 2t + 1$, some $t \geq 1$.

Then n is the lcm of the orders of

(1) $\alpha, \alpha^2, \alpha^3, \ldots, \alpha^{2t}$.

It is this subset of BCH codes that we will deal with exclusively in the remainder of this chapter.

Example 4

Let α be a root of $x^4 + x + 1$ over GF(2). Then α is a primitive element of $GF(2^4)$. Let $t = 3$ and consider the BCH code of all binary polynomials $f(x)$ of degree $\leq n - 1$ having

$$\alpha, \alpha^2, \alpha^3, \alpha^4, \alpha^5, \text{ and } \alpha^6 \text{ as roots.}$$

Here, n = 15.

Now, if $f(x)$ is in $Z_2[x]$ and β is any element of $GF(2^m)$ then

$$(f(\beta))^2 = f(\beta^2).$$

So,

$$f(\alpha^2) = f(\alpha)^2$$

$$f(\alpha^4) = f(\alpha^2)^2 = f(\alpha)^4$$

$$f(\alpha^6) = f(\alpha^3)^2.$$

Thus, the same set of polynomials results if we require only that α, α^3, α^5 be roots. But now we are in the situation of Example 2 and the resulting BCH code is a (15, 5) code.

Remarks:

1) Generalizing the result of the above Example, the same code results from Definition (1) if we require only that

$$\alpha, \alpha^3, \alpha^5, \ldots, \alpha^{2t-1}$$

be roots of the polynomials $f(x)$.

2) Since α has order $2^m - 1$ and any power of α has order dividing $2^m - 1$, it follows that in (1), it is always the case that $n = 2^m - 1$.

3) If $2t >$ order of $\alpha = n$, then the list of roots in (1) contains duplication. So it suffices to consider the cases $t \leq \dfrac{n}{2}$

Example 5

Let α be a primitive element of $GF(2^4)$ with minimim polynomial $x^4 + x + 1$ over $GF(2)$. Let $t = 2$.

Then code vectors are polynomials of degree ≤ 14 having α and α^3 as roots. Using the results of Example 2,

$$g(x) = \text{lcm}(x^4 + x + 1 , x^4 + x^3 + x^2 + x + 1)$$

$$= 1 + x^4 + x^6 + x^7 + x^8$$

is a generator polynomial.
This gives a (15, 7) BCH code.

Error Correction Capabilities

We now aim to show that the BCH codes provide us with a large class of multiple error-correcting codes.

Proposition

If $t \leq \frac{n}{2}$, the BCH code of (1) has minimum distance at least d.

Proof:

The code is the null space of the matrix

$$
H = \begin{pmatrix}
1 & \alpha & \alpha^2 & . & . & . & \alpha^{n-1} \\
1 & \alpha^2 & (\alpha^2)^2 & . & . & . & (\alpha^2)^{n-1} \\
1 & \alpha^3 & (\alpha^3)^2 & . & . & . & (\alpha^3)^{n-1} \\
. & . & . & & & & . \\
. & . & . & & & & . \\
. & . & . & & & & . \\
1 & \alpha^{2t} & (\alpha^{2t})^2 & . & . & . & (\alpha^{2t})^{n-1}
\end{pmatrix}
$$

$$2t \times n$$

Consider any set of $d-1 = 2t$ columns of H and the determinant of the square matrix they form.

$$
\begin{vmatrix}
\alpha^{j_1} & . & . & . & \alpha^{j_{2t}} \\
(\alpha^{j_1})^2 & . & . & . & (\alpha^{j_{2t}})^2 \\
. & & & & . \\
. & . & . & . & . \\
. & & & & . \\
(\alpha^{j_1})^{2t} & . & . & . & (\alpha^{j_{2t}})^{2t}
\end{vmatrix}
$$

59

Factoring an α^{j_i} out of each column, the determinant becomes

$$\alpha^{j_1} \ldots \alpha^{j_{2t}} \begin{vmatrix} 1 & 1 & \cdot & \cdot & \cdot & 1 \\ \alpha^{j_1} & \alpha^{j_2} & \cdot & \cdot & \cdot & \alpha^{j_{2t}} \\ \cdot & \cdot & & & & \cdot \\ \cdot & \cdot & & & & \cdot \\ \cdot & \cdot & & & & \cdot \\ (\alpha^{j_1})^{2t-1} & (\alpha^{j_2})^{2t-1} & \cdot & \cdot & \cdot & (\alpha^{j_{2t}})^{2t-1} \end{vmatrix}$$

But this is a Vandermonde determinant whose value is

$$\alpha^{j_1} \ldots \alpha^{j_{2t}} \prod_{j_i > j_k} (\alpha^{j_i} - \alpha^{j_k}).$$

Since no two powers of α in the list $\alpha, \alpha^2, \ldots, \alpha^{2t}$ are equal (here we use $2t \leq n$) this determinant is non-zero. Thus, no set of $d - 1$ columns of H is linearly dependent. Proposition 2 of Chapter 3 gives our result.

Put another way, this says

Proposition:

If $t \leq \frac{n}{2}$, the BCH code described in (1) is capable of correcting all combinations of t or fewer errors.

Proof:

Note that $d = 2t + 1$. Use Proposition 3 of Chapter 3.

Remarks:

Thus, the (15, 5) BCH code of Example 4 is triple error-correcting, while the (15, 7) BCH code of Example 5 is double error-correcting. Peterson and Weldon [7], p. 274, have a table of BCH codes with $n \leq 2^{10}$ along with their error correcting capabilities. Note that the above proposition does not provide a constructive error correcting procedure. It does allow the receiver to compare a received word with all 2^k possible codewords of an (n, k) code and pick the closest codeword as the sent message. This is a fairly formidable task for even moderate values of k -- a (31, 21) BCH code would involve searching through more than 1 million codewords. In the next section we outline a decoding procedure that aims at much greater efficiency.

A Decoding Procedure

Fix a BCH code of the type described in (1). We wish to describe a procedure that allows the correction of any set of t or fewer errors.

Suppose the codeword c(x) is sent (we adopt the polynomial viewpoint), while r(x) is received.
Then

$$r(x) = c(x) + e(x), \text{ for some } e(x).$$

e(x) is the error polynomial (or n-tuple) that has 1's in those positions that have been altered.

More precisely, suppose that v actual errors have occurred, $v \leq t$, in the positions

$$m_1, m_2, \ldots, m_v.$$

Then

$$e(x) = x^{m_1} + x^{m_2} + \ldots + x^{m_v}.$$

Note that at this stage the receiver does not know the value of v.

If e(x) were known, then all is well since
$$c(x) = r(x) + e(x) \qquad \text{(all polynomials are over GF(2)).}$$

Now, $c(\alpha^i) = 0$, for i = 1, . . ., 2t, by the definition of the code.
So,

$$r(\alpha^1) = c(\alpha^i) + e(\alpha^i)$$
$$= e(\alpha^i), \qquad\qquad i = 1, \ldots, 2t.$$

Since r(x) is known, the elements

$$e(\alpha^i) = (\alpha^{m_1})^i + (\alpha^{m_2})^i + \ldots + (\alpha^{m_v})^i$$

of $GF(2^m)$ are known.

Some notation:

Let

$$Y_1 = \alpha^{m_1}$$
$$Y_2 = \alpha^{m_2}$$
$$\vdots \qquad \vdots$$
$$Y_v = \alpha^{m_v}$$

and

let
$$S_1 = \sum_{j=1}^{v} Y_j^i , \qquad i = 1, \ldots, 2t.$$

So,
$$S_i = e(\alpha^i), \quad i = 1, \ldots, 2t.$$

Thus, <u>we know</u>

$$S_i, \; i = 1, \ldots, 2t;$$

<u>our problem</u>: find v

and
$$Y_i, \; i = 1, \ldots, v.$$

Note that expressing the Y_i as powers of the primitive element α would give the error locations $m_1, \ldots, m_v$.

Our approach to finding the Y_i is to construct a polynomial for which the Y_i are the roots.

Certainly the polynomial
$$(Z + Y_1) (Z + Y_2) \ldots (Z + Y_v)$$

with coefficients in $GF(2^m)$ has exactly the roots $Y_i, \; i = 1, \ldots v.$

The elementary theory of equations (or brute force multiplication) says

$$(Z + Y_1) \ldots (Z + Y_v) \tag{2}$$

$$= Z^v + \sigma_1 Z^{v-1} + \sigma_2 Z^{v-2} + \ldots + \sigma_{v-1} Z + \sigma_v$$

where

$$\sigma_1 = Y_1 + Y_2 + \ldots + Y_v$$

$$\sigma_2 = \sum_{1 \le i \le j \le n} Y_i Y_j$$

.
.
.

$$\sigma_v = Y_1 Y_2 Y_3 \ldots Y_v$$

(the σ_i are the <u>elementary symmetric functions</u> of $Y_1, \ldots, Y_v$).

(2) is called the <u>error locator polynomial</u>.

<u>The point is:</u> the S_i and the σ_i are related via what are sometimes called Newton's identities. We include a short derivation.

Multiply both sides of (2) by Y_j^i and let $Z = Y_j$, to get

$$0 = Y_j^i \left[Y_j^v + \sigma_1 Y_j^{v-1} + \sigma_2 Y_j^{v-2} + \ldots + \sigma_{v-1} Y_j + \sigma_v \right]$$

<u>i.e.,</u>

$$0 = Y_j^{i+v} + \sigma_1 Y_j^{i+v-1} + \sigma_2 Y_j^{i+v-2} + \ldots + \sigma_{v-1} Y_j^{i+1} + \sigma_v Y_j^i.$$

Sum the above equations for $1 \le j \le v$, producing

$$S_i \sigma_v + S_{i+1} \sigma_{v-1} + \ldots + S_{i+v-1} \sigma_1 + S_{i+v} = 0 \tag{3}$$

for $i = 1, \ldots, 2t.$

Consider the v equations in the v unknowns $\sigma_v, \sigma_{v-1}, \ldots, \sigma_1$ obtained by limiting i in the above to range between 1 and v.

In matrix language, this system can be written as

$$\begin{pmatrix} S_1 & S_2 & \cdot & \cdot & \cdot & S_v \\ S_2 & S_3 & \cdot & \cdot & \cdot & S_{v+1} \\ \cdot & \cdot & \cdot & \cdot & \cdot & \cdot \\ S_v & S_{v+1} & \cdot & \cdot & \cdot & S_{2v-1} \end{pmatrix} \begin{pmatrix} \sigma_v \\ \sigma_{v-1} \\ \cdot \\ \cdot \\ \cdot \\ \sigma_1 \end{pmatrix} = \begin{pmatrix} S_{v+1} \\ S_{v+2} \\ \cdot \\ \cdot \\ \cdot \\ S_{2v} \end{pmatrix} \qquad (4)$$

where v = number of errors that have occured.

If the above square matrix were invertible, we could then use it to solve for the σ_i in terms of the known S_i. We would then know the coefficients of the error locator polynomial (2) and by systematically plugging in the non-zero elements of the field $GF(2^m)$ we could isolate its roots -- i.e. find the Y_i and thus know the error locations m_i.

Aside of the invertibility of the matrix in (4), one other problem still remains:

What is v?

The answer is:

v is the largest $\ell \leq t$ for which the $\ell \times \ell$ matrix $\qquad$ (5)

$$\begin{pmatrix} S_1 & S_2 & \cdot & \cdot & \cdot & S_\ell \\ S_2 & S_3 & \cdot & \cdot & \cdot & S_{\ell+1} \\ \cdot \\ \cdot \\ \cdot \\ S_\ell & & \cdot & \cdot & \cdot & S_{2\ell-1} \end{pmatrix}$$

is invertible

A proof of this claim is deferred to the end of this section.

We now have at hand a <u>decoding procedure</u> which we <u>summarize</u> below and illustrate by several examples.

1. Compute S_i, $i = 1, \ldots, 2t$

   Note: $S_i = r(\alpha^i)$

2. Find v

   -- for example, by taking the determinants

$$\begin{vmatrix} S_1 & S_2 & \cdot & \cdot & \cdot & S_\ell \\ \cdot & & & & & \cdot \\ \cdot & & & & & \cdot \\ \cdot & & & & & \cdot \\ S_\ell & \cdot & \cdot & \cdot & \cdot & S_{2\ell-1} \end{vmatrix}$$

   working down from $\ell = t$ to $\ell = 1$. v is then the first value of ℓ for which the determinant is nonzero.

3. Solve equation (4) for $\sigma_1, \sigma_2, \ldots, \sigma_v$, thus getting the coefficients of the error locator polynomial $p(Z)$.

4. Find the v roots Y_i of $p(Z)$. Write $Y_i = \alpha^{m_i}$ for some m_i.

5. Then $m_1, m_2, \ldots, m_v$ give the error locations.

<u>Example 6</u>

Consider the (15, 5) BCH code of Example 4. Here, $t = 3$. Suppose the received vector is

$$1\ 1\ 1\ 1\ 1\ 1\ 1\ 0\ 0\ 1\ 0\ 1\ 0\ 0\ 0$$

So that

$$r(x) = 1 + x + x^2 + x^3 + x^4 + x^5 + x^6 + x^9 + x^{11}$$

then

$$S_1 = r(\alpha) = 1 + \alpha + \alpha^2 + \alpha^3 + \alpha^4 + \alpha^5 + \alpha^6 + \alpha^9 + \alpha^{11}$$

$$= 1 + \alpha + \alpha^2 + \alpha^3 + \alpha + 1 + \alpha^2 + \alpha + \alpha^3 + \alpha^2 + \alpha^3 + \alpha + \alpha^3 + \alpha^2 + \alpha$$

$$= \alpha \qquad \text{(Table 2 is a must)}$$

65

$$S_2 = r(\alpha^2) = r(\alpha)^2 = S_1^2 = \alpha^2$$

$$\text{(note that in general } S_{2i} = (S_i)^2)$$

$$S_3 = r(\alpha^3) = 1 + \alpha^3 + \alpha^6 + \alpha^9 + \alpha^{12} + \alpha^{15} + \alpha^{18} + \alpha^{27} + \alpha^{33}$$

$$= 1 + \alpha^3 + \alpha^6 + \alpha^9 + \alpha^{12} + 1 + \alpha^3 + \alpha^{12} + \alpha^3$$

$$= \alpha^3 + \alpha^6 + \alpha^9 = \alpha^3 + \alpha^3 + \alpha^2 + \alpha^3 + \alpha$$

$$= \alpha + \alpha^2 + \alpha^3 = \alpha^{11}$$

$$S_4 = (S_2)^2 = \alpha^4$$

$$S_5 = 1 + \alpha^5 + \alpha^{10} + \alpha^{15} + \alpha^{20} + \alpha^{25} + \alpha^{30} + \alpha^{45} + \alpha^{55}$$

$$= 1 + \alpha^5 + \alpha^{10} + 1 + \alpha^5 + \alpha^{10} + 1 + 1 + \alpha^{10}$$

$$= \alpha^{10}$$

$$S_6 = (S_3)^2 = \alpha^{22} = \alpha^7$$

We now find v:

$$\begin{vmatrix} S_1 & S_2 & S_3 \\ S_2 & S_3 & S_4 \\ S_3 & S_4 & S_5 \end{vmatrix} = \begin{vmatrix} \alpha & \alpha^2 & \alpha^{11} \\ \alpha^2 & \alpha^{11} & \alpha^4 \\ \alpha^{11} & \alpha^4 & \alpha^{10} \end{vmatrix}$$

$$= \alpha \cdot \alpha^2 \cdot \alpha^4 \begin{vmatrix} 1 & \alpha & \alpha^{10} \\ 1 & \alpha^9 & \alpha^2 \\ \alpha^7 & 1 & \alpha^6 \end{vmatrix} \quad \text{(expanding by the 1}^{\text{st}}\text{ column)}$$

$$= \alpha^7 \left[\begin{vmatrix} \alpha^9 & \alpha^2 \\ 1 & \alpha^6 \end{vmatrix} + \begin{vmatrix} \alpha & \alpha^{10} \\ 1 & \alpha^6 \end{vmatrix} + \alpha^7 \begin{vmatrix} \alpha & \alpha^{10} \\ \alpha^9 & \alpha^2 \end{vmatrix} \right]$$

$$= \alpha^7 \left[\alpha^{15} + \alpha^2 + \alpha^7 + \alpha^{10} + \alpha^{10} + \alpha^{26} \right]$$

$$= \alpha^7 \left[1 + \alpha^2 + \alpha^3 + \alpha + 1 + \alpha^3 + \alpha^2 + \alpha \right]$$

$$= \alpha^7 \cdot 0 = 0$$

So, $v \leq 2$ (two or fewer errors have occured).

$$\begin{vmatrix} S_1 & S_2 \\ S_2 & S_3 \end{vmatrix} = \begin{vmatrix} \alpha & \alpha^2 \\ \alpha^2 & \alpha^{11} \end{vmatrix} = \alpha^{12} + \alpha^4$$

$$= \alpha^3 + \alpha^2 + \alpha + 1 + \alpha + 1 = \alpha^3 + \alpha^2 = \alpha^6 \neq 0.$$

So, $v = 2$.

Equations (4) then become

$$\alpha \sigma_2 + \alpha^2 \sigma_1 = \alpha^{11}$$

$$\alpha^2 \sigma_2 + \alpha^{11} \sigma_1 = \alpha^4$$

or

$$\sigma_2 + \alpha \sigma_1 = \alpha^{10}$$

$$\sigma_2 + \alpha^9 \sigma_1 = \alpha^2$$

Thus

$$\sigma_1 = \frac{\alpha^{10} + \alpha^2}{\alpha + \alpha^9} = \frac{\alpha^2 (1 + \alpha^8)}{\alpha (1 + \alpha^8)} = \alpha$$

and

$$\sigma_2 = \alpha^{10} + \alpha \sigma_1 = \alpha^{10} + \alpha^2 = \alpha + 1 = \alpha^4.$$

The error locator polynomial is then

$$Z^2 + \alpha Z + \alpha^4 = (Z + \alpha^4)(Z + 1).$$

Its roots can be found by evaluating it at

$Z = \alpha^i$, $i = 0, \ldots, 14$, or by an on the spot factorization.

67

They are α^4 and $1 = \alpha^0$, and so the error positions are

$$m_1 = 0 \quad \text{and} \quad m_2 = 4.$$

The received word is then decoded as the codeword

$$0\ 1\ 1\ 1\ 0\ 1\ 1\ 0\ 0\ 1\ 0\ 1\ 0\ 0\ 0.$$

Note that this codeword is $xg(x)$, where $g(x)$ is the generator polynomial of the code.

Example 7

Here we use the (15, 7) BCH code of Example 5, where $t = 2$. We omit some of the intermediate arithmetic. Suppose the received word is

$$1\ 0\ 0\ 0\ 0\ 0\ 1\ 1\ 1\ 1\ 0\ 0\ 0\ 0\ 0,$$

so that

$$r(x) = 1 + x^6 + x^7 + x^8 + x^9$$

$$S_1 = r(\alpha) = \alpha^{14}$$

$$S_2 = S_1^2 = \alpha^{13}$$

$$S_3 = r(\alpha^3) = 0$$

$$S_4 = S_2^2 = \alpha^{11}$$

$$\begin{vmatrix} S_1 & S_2 \\ S_2 & S_3 \end{vmatrix} = \alpha^{26} \neq 0,$$

so,

$$v = 2.$$

Equations (4) become

$$\alpha^{14}\sigma_2 + \alpha^{13}\sigma_1 = 0$$

$$\alpha^{13}\sigma_2 = \alpha^{11}$$

68

with solutions

$$\sigma_2 = \alpha^{13}, \quad \sigma_1 = \alpha^{14}.$$

The error locator polynomial is

$$Z^2 + \alpha^{14} Z + \alpha^{13},$$

whose roots are α^4 and α^9.

Thus, errors have occured in positions 4 and 9 and we decode as

$$1\ 0\ 0\ 0\ 1\ 0\ 1\ 1\ 1\ 0\ 0\ 0\ 0\ 0\ 0 \ .$$

Verification of the Claim (5)

Let $v \leq \ell \leq t$, where v = number of errors that have occured.

Let $Y_j = 0$ for $v < j \leq \ell$.

Then we can write

$$S_i = \sum_{j=1}^{\ell} Y_j^i .$$

The following is a matrix identity

$$\begin{pmatrix} S_1 & S_2 & \cdot & \cdot & \cdot & S_\ell \\ S_2 & & & & & S_{\ell+1} \\ \cdot & & & & & \\ \cdot & & & & & \\ S_\ell & & & & & S_{2\ell-1} \end{pmatrix} =$$

$$\begin{pmatrix} 1 & 1 & \cdot & \cdot & \cdot & 1 \\ Y_1 & Y_2 & \cdot & \cdot & \cdot & Y_\ell \\ & & & & & \cdot \\ & & & & & \cdot \\ Y_1^{\ell-1} & Y_2^{\ell-1} & & & & Y_\ell^{\ell-1} \end{pmatrix} \begin{pmatrix} Y_1 & & & & 0 \\ & Y_2 & & & \\ & & \cdot & & \\ & & & \cdot & \\ 0 & & & & Y_\ell \end{pmatrix} \begin{pmatrix} 1 & Y_1 & Y_1^2 & \cdot & \cdot & \cdot & Y_1^{\ell-1} \\ 1 & Y_2 & \cdot & \cdot & \cdot & \cdot & Y_2^{\ell-1} \\ \cdot & & & & & & \\ \cdot & & & & & & \\ 1 & Y_\ell & \cdot & \cdot & \cdot & \cdot & Y_\ell^{\ell-1} \end{pmatrix}$$

$$= ADA^T$$

69

which can be directly verified by multiplying out the right hand side.

If $\ell > v$, then D has a zero on the diagonal and is thus singular, as is then the matrix on the left.

If $\ell = v$, then each Y_j is non-zero, so that D is invertible. Further, A and A^T are Vandermonde matrices with

$$|A| = |A^T| = \prod_{v \geq i \geq j \geq 1} (Y_i - Y_j)$$

By construction the Y_i are distinct since $Y_i = \alpha^{m_i}$, where the m_i are the error locations. So, $|A| \neq 0$, and thus all matrices in (6) are invertible.

Concluding Remarks:

While the decoding algorithm we have described is far superior to the brute checking of all codewords, for large t it does require evaluation of sizeable determinants over $GF(2^m)$. There is an alogrithm due to Berlekamp [1] that is computationally less complex.

The references [1]. [6], and [7] contain a wealth of material on algebraic coding. They also contain discussions of the circuitry necessary to implement the arithmetic of finite fields used in coding and decoding.

[2] is a collection of papers that have significantly influenced the development of coding theory. Many are readily accessible to the student with a first course in abstract algebra and make interesting reading that can impart a bit of historical perspective.

References

[1]   E. R. Berlekamp,   Algebraic Coding Theory, McGraw-Hill, New York, 1968.

[2]   Ian F. Blake,   Algebraic Coding Theory: History and
      editor   Development, Dowden, Hutchinson & Ross, Stroudsbourg, Pennsylvania, 1973.

[3]   Ian F. Blake,   "Codes and Designs", Mathematics Magazine,
                                 Vol. 52, No. 2, March 1979, p. 81-95.

[4]   R. W. Hamming,   Coding and Information Theory, Prentice-
                                 Hall, Englewood Cliffs, New Jersey, 1980.

[5]   N. Levinson,   "Coding Theory: A Counterexample to G. H.
                                 Hardy's Conception of Applied Mathematics".
                                 American Mathematical Monthly, Vol. 77, No.
                                 3, March 1970, p. 249-258.

[6]   F. J. MacWilliams   The Theory of Error-Correcting Codes,
      and N.J.A. Sloane   North-Holland, Amsterdam, 1977.

[7]   W W. Peterson and   Error-Correcting Codes (Second Edition)
      E. J. Weldon, Jr.   MIT Press, Cambridge, 1972.

[8]   Vera Pless,   Intorduction to the Theory of Error-
                                 Correcting Codes, John Wiley and Sons,
                                 1982.

1. Let α be as in Table 2.
   Write out the 4×15 matrix $(1 \ \alpha \ \alpha^2 \ \ldots \ \alpha^{14})$ and note that up to a permutation of columns, it is the parity check matrix of the $(15, 11)$ Hamming code.

2. Verify the computations in Table 1.

3. If α is defined as in Table 2, verify that $\alpha^9 = \alpha^3 + \alpha$.

4. List two non-zero codewords in the $(15,12)$ code of Example 1.

5. Let α be the primitive element of Table 1. Consider the code consisting of all $f(x)$ of $\deg \leq n - 1$ with $f(\alpha) = f(\alpha^3) = f(\alpha^5) = 0$. Find a generator polynomial for this code and list all the codewords in the code.

6. Construct a $(15, 10)$ cyclic code over $GF(2)$.

7. Suppose $(c_0, c_1, \ldots, c_{n-1})$ is a code word in a cyclic code with generator polynomial $g(x)$. Show that $(c_{n-1}, c_0, c_1, \ldots, c_{n-2})$ is also a codeword.

   Hint:

   $$c_{n-1} + c_0 x + \ldots + c_{n-2} x^{n-1}$$
   $$= x(c_0 + c_1 x + \ldots + c_{n-1} x^{n-1}) - c_{n-1}(x^n - 1).$$

8. Using the $(15, 5)$ BCH code of Example 4 decode the received message

   $$1\ 1\ 0\ 0\ 0\ 1\ 0\ 0\ 1\ 1\ 0\ 1\ 0\ 0\ 0$$

9. Using the $(15, 7)$ BCH code of Example 5 decode the received message

   a) $\quad 1\ 0\ 1\ 0\ 1\ 0\ 0\ 1\ 0\ 0\ 0\ 0\ 0\ 0\ 0$

   b) $\quad 0\ 0\ 1\ 0\ 0\ 0\ 1\ 0\ 1\ 1\ 1\ 0\ 0\ 0\ 0$

72

10.   a) Show that the case $t = 1$ of Def. (1) yields a Hamming code.

      b) When $t = 1$ show that the error correction procedure for BCH codes reduces to that used for Hamming codes in Chapter 3.

5   Crystallographic Groups in the Plane I

Results concerning symmetry find a natural expression in the language of group theory and one of the most striking examples of this is found in crystallography, the study of crystals. The distinguishing characteristic of a crystal is that it is based on the repetition of some basic motif. Crystals can be thought of as 3-dimensional analogues of repeating wallpaper patterns -- basic building blocks, clusters of atoms, are stacked together in a regular pattern in space to form the crystal.

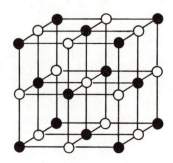

It is the geometry of this pattern and the symmetries it permits that concern crystallographers and form the basis of their classification of crystals. By making these intuitive notions more precise through the use of mathematics, they have shown that crystals can exhibit only one of 230 basic types of symmetries.

While not difficult, the arguments used in classifying crystals in 3 dimensions can be a bit long-winded. The essential ideas involved can be adequately displayed in 2 dimensions, with less risk of inundating a reader with numbing detail. We will thus confine ourselves to the study of symmetries of 2-dimensional crystals. In the process we will demonstrate that there are only 17 types of distinct repeating 'wallpaper patterns' that can exist in the plane.

A brief overview may be in order. Crystal classification is accomplished by assigning a group (called a space group) to each crystal, the group consisting of the symmetries that the crystal exhibits. Two crystals are considered different if their space groups are distinct. In this way sorting crystals according to symmetry type becomes the group theoretic problem of asking, How many distinct space groups are possible? The mathematical analysis begins by showing that any space group is composed of two parts: an infinite abelian subgroup consisting of the translational symmetries of the crystal (the translation subgroup) and a finite group consisting roughly of the rotations and reflections the crystal allows (the point group). An elegant elementary argument is used to show that, in two dimensions, the point group must be one of only ten possible groups. Also, geometry is used to limit the possibilities for the translation subgroup to finitely many. The final classification is accomplished by examining the number of ways these finitely many groups can combine to produce a crystallographic space group.

Our discussion is divided into two parts. In this chapter we introduce space groups, give examples, and indicate how point groups and translation subgroups are related to space groups. We also limit point groups to finitely many possibilities. Since 2 x 2 orthogonal matrices play a role in the arguments, a short appendix on these follows this chapter. The succeeding chapter examines the structure of the translation subgroup and pursues in detail how the ten possible point groups can interract with translation subgroups to form space groups. The end result of the analysis is the construction of the 17 possible distinct space groups in two dimensions.

Lattices

The idea that a crystal is formed by stacking together identical copies of a basic cell is modeled in 3 dimensions by considering copies of a parallelepiped that fill space.

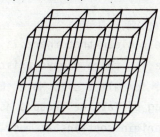

The two dimensional version of this involves the repetition of a fixed parallelogram.

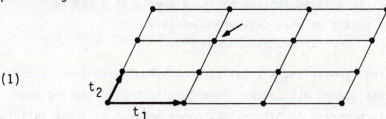

(1)

The vertices of these parallelograms form a regular array of points in the plane, called a lattice. More precisely,

__Def.__   Let t_1, t_2 be linearly independent vectors in R^2 (the plane).

A __lattice__ is the set T of all vectors of the form.

$$t = mt_1 + nt_2,$$

where m and n are integers.

t_1, t_2 are called a __lattice basis__ of T.

__Lattice vectors__ are elements of T.

Remarks & Examples

1. In (1), the vectors t_1, t_2 form a lattice basis. The indicated point corresponds to $t_1 + 2t_2$.

2. A lattice T may have many lattice bases. For example, the lattice with basis (1, 0) and (0, 1)

(2) 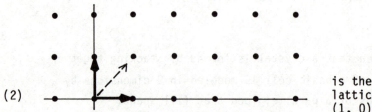 is the same as the lattice with basis (1, 0) and (1, 1)

3. It should be intuitively clear that there is a minimum length to the vectors in any lattice T: there always exists a non-zero $t_0 \in T$ with $||t_0|| \leq ||t||$, for all $t \in T$. Note that t_0 may not be unique. Also, thinking of lattice vectors as points in the plane, any circle about the origin contains at most finitely many lattice vectors.

4. With respect to a fixed lattice basis, elements of T can be represented by tuples with integer coordinates:

(3) $$mt_1 + nt_2 \longleftrightarrow \binom{m}{n}$$

5. In discussing symmetries, crystallographers deal with idealized crystals that occupy all space. They justify this by noting that, for crystals occuring in nature, distances between adjacent building blocks are on an atomic level and by arguing that the external symmetry of a real crystal shown by the arrangement of its faces is derived from symmetries of the corresponding idealized version.

We should note that the representation in (3) depends on the lattice basis chosen. How coordinates change as bases change is easily seen.

Suppose t_1, t_2 and s_1, s_2 are both lattice bases for T. Then,

$$t_1 = \alpha s_1 + \beta s_2$$
$$t_2 = \gamma s_1 + \delta s_2 \qquad \text{for some \underline{integers}}$$

α, β, γ, δ. The matrix $U = \begin{pmatrix} \alpha & \gamma \\ \beta & \delta \end{pmatrix}$

76

has the property that $U\binom{m}{n}$ gives the coordinates of the lattice vector $t = mt_1 + nt_2$ w.r.t. s_1, s_2.

Also, $U^{-1}\binom{k}{\ell}$ gives the co-ordinates of the lattice vector $s = ks_1 + \ell s_2$ with respect to t_1, t_2.

It follows that
U^{-1} must also have integer entries.

Since
$$UU^{-1} = \begin{pmatrix} 1 & 0 \\ 0 & 1 \end{pmatrix}$$

we have
$$(\det U)(\det U^{-1}) = 1.$$

But det U and det U^{-1} are both <u>integers</u>. Thus, det $U = 1$ or -1.

<u>Def.</u> A square matrix with <u>integer</u> entries whose determinant is +1 or
       -1 is called <u>unimodular</u>.

We have shown that changing lattice bases is effected by multiplying by a unimodular matrix.

<u>Example</u>:   in (2) above,

$$U = \begin{pmatrix} 1 & -1 \\ 0 & 1 \end{pmatrix}$$

<u>Symmetries</u>

Intuitively, a symmetry of an object is a transformation of the object that leaves the object looking the same. We make this more precise by first discussing symmetries in general and then specializing to crystals.

<u>Def.</u>   A <u>symmetry</u> of R^2 is a 1 - 1 and onto mapping
       $f: R^2 \to R^2$ that preserves distances: that is,
$$||f(x) - f(y)|| = ||x - y||,$$
       for all $x, y \in R^2$.

Examples:

1. the identity mapping

2. translation by a vector a ε R^2 -- i.e. the mapping
   $T_a(x) = x + a$, $\forall$ x ε R^2.

3. rotation through an angle θ .

The set of all symmetries of R^2 forms a group under composition -- we denote this group by E(2) (E is for 'Euclidean'). The set of all translations $\{T_a | a \varepsilon R^2\}$ forms an infinite abelian subgroup of E(2). Note that the inverse of T_a is T_{-a}. Likewise, the set of all rotations forms a subgroup of E(2).

What does a typical element f of E(2) look like? Let f(0) = a, where 0 denotes the origin of R^2. Then the element $T_{-a}f$ of E(2) fixes the origin:

$$T_{-a}f(0) = T_{-a}(a) = a - a = 0.$$

But a symmetry which fixes the origin must be a linear transformation (see Exercise 12). Thus, there exists a distance preserving linear transformation A with $T_{-a}f = A$.
So,

$$f = T_aA.$$

That is, for any vector x in R^2,

(4) $f(x) = T_aA(x) = Ax + a.$

So, any symmetry is a distance preserving linear transformation followed by a translation.

Let f and g be elements of E(2) with

$$f(x) = Ax + a$$

and

$$g(x) = Bx + b.$$

Then the composite fg satisfies

$$
\begin{aligned}
(fg)(x) &= f(g(x)) \\
&= f(Bx + b) \\
&= A(Bx + b) + a \\
&= ABx + (Ab + a).
\end{aligned}
$$

78

Thus, in light of (4), if we think of elements of E(2) as tuples of the form (A, a), the group multiplication in E(2) is given by

  1) $(A, a)(B, b) = (AB, Ab + a)$.

Also,

  2)   The identity of E(2) is $(I, 0)$, where I is the identity
(5)   mapping.

  3)   The inverse of the element (A, a) is given by
       $(A, a)^{-1} = (A^{-1}, - A^{-1}a)$

  4) T_a is represented by (I, a).

It will be to our advantage to write symmetries in the form (A, a).

A further word is in order on expressing elements of E(2) that hinges on the correspondence between linear transformations and matrices.

An element of E(2) can be thought of 'abstractly' as a linear transformation, A, coupled with a translation vector, a. Fixing a basis of R^2, we can compute the matrix of the linear transformation and the co-ordinates of the translation vector. Thus, A can be thought of as a matrix, and a as a column vector of co-ordinates. But this representation is basis dependent.

Much like in our discussion of change of lattice bases, if we change bases of R^2 the column vector a now becomes Ca and the matrix A becomes the matrix CAC^{-1}, where C is a change of basis matrix.

Example: reflection through the x-axis has matrix $\begin{pmatrix} 1 & 0 \\ 0 & -1 \end{pmatrix}$
         with respect to the basis (1, 0), (0, 1); its matrix w.r.t.
         to the basis (1, 0), (2, 1) is $\begin{pmatrix} 1 & 4 \\ 0 & -1 \end{pmatrix}$. Here, $C = \begin{pmatrix} 1 & -2 \\ 0 & 1 \end{pmatrix}$.
         Thus, the element ($\begin{pmatrix} 1 & 0 \\ 0 & -1 \end{pmatrix}$, $\begin{pmatrix} 1 \\ 1 \end{pmatrix}$) of E(2), after changing
         bases is written as
         $$\left(\begin{pmatrix} 1 & 4 \\ 0 & -1 \end{pmatrix}, \begin{pmatrix} -1 \\ 1 \end{pmatrix} \right).$$

         It will often be convenient for us to write elements of E(2)
         in such 'matrix form'. We will take care to indicate what
         the underlying basis is.

Symmetries of Crystals: Space Groups

We wish to examine symmetry groups of crystals -- namely, those symmetries which map a crystal to itself. The intuitive notion that a crystal is formed of repetition of fundamental building blocks is expressed by saying that there is some lattice T with basis t_1, t_2 such that any translation of the form T_a , where a = $mt_1 + nt_2$, is a symmetry of the crystal. Further, to denote the fact that a crystal is formed of discrete units, we insist that any translation which maps the crystal onto itself arise from the lattice T -- i.e. it must be of the form T_a, where a ϵ T. This is equivalent to insisting that, among all translations which map a crystal onto itself, there be a translation of minimum length. Now, besides those translations, there may be other elements of E(2) that map the crystal onto itself -- these would also be elements of the symmetry group of the crystal. Crystallographers classify crystals according to the symmetry groups that the crystals have. Pursuing such a classification is tantamount then to finding those subgroups of E(2) whose translations are generated by a lattice.

Def.   A crystallographic space group (space group) is a subgroup G of E(2) whose translations are a set of the form {(I, t) | t ϵ T} where T is a lattice.

Remarks:

1. Any space group is necessarily infinite -- it contains at least infinitely many translations.
2. The set of translations {(I, t) |t ϵ T} forms an abelian subgroup of G, called the translation subgroup of G. Clearly there is a 1 - 1 correspondence

$$(I, t) \longleftrightarrow t$$

between translations in G and elements of the lattice T. Thus, no confusion should result if we use the letter T to denote both the translation subgroup of G and the associated lattice.

Examples:

Let T be the lattice with basis (1, 0) and (0, 1). The matrices and vectors below are written w.r.t. this lattice basis.

1) G = T is a space group.
2) Let G be the set consisting of the translation subgroup T along with all elements of the form

80

$$(A, t), \ t \ \epsilon \ T,$$

where
$$A = \begin{pmatrix} -1 & 0 \\ 0 & -1 \end{pmatrix}.$$

(6) $\underline{E.g.}$, $(I, \begin{pmatrix} 2 \\ 3 \end{pmatrix})$ and $(\begin{pmatrix} -1 & 0 \\ 0 & -1 \end{pmatrix}, \ \begin{pmatrix} 1 \\ 1 \end{pmatrix})$ are elements of G. Then G is a space group. (Check that G is indeed a group - $\underline{e.g.}$, closed under multiplication).

3) Let G be the set consisting of T along with all elements of the form $(A, t), \ t \ \epsilon \ T$ where

(7) $$A = \begin{pmatrix} 1 & 0 \\ 0 & -1 \end{pmatrix}.$$ Then G is a space group.

4) Let G consist of T along with all elements of the form

(8) $(A, t + (\begin{smallmatrix} 1/2 \\ 0 \end{smallmatrix}))$, where $A = \begin{pmatrix} 1 & 0 \\ 0 & -1 \end{pmatrix}$, $t \ \epsilon \ T$. G is a space group.

Below are pictures corresponding to the above space groups. In each case a parallelogram whose repetition forms the lattice is shown. Symbols are inserted to indicate the non-translational symmetries the group contains. Thus, 2) permits a 180° rotation, while 3) does not; 3) permits reflection through the x-axis, while 2) does not.

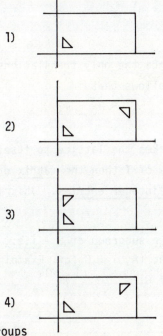

1)

2)

3)

4)

Note that the lattice itself does not identify the crystal by symmetry type. Intuitively, think of the crystal as having identical patterns of atoms at each lattice point. The lattice provides the translational symmetry of the crystal; the type of atom pattern determines the full symmetry group. Note also that 4) shows that a space group may contain elements of the form (A, a), where a is not a lattice vector.

Point Groups

A space group G, being a subgroup of E(2), consists of elements of

81

the form (A, a) where A is a linear transformation and a is a vector.

Def. The point group G_0 of G is the set

$$G_0 = \{A \mid (A, a) \in G, \text{ for some } a\}.$$

Remarks & Examples

1. G_0 consists of all possible 'linear parts' of symmetries that are in G. The relations (5) guarantee that G_0 is indeed a group.

2. In example (6),
$$G_0 = \{I, \begin{pmatrix} -1 & 0 \\ 0 & -1 \end{pmatrix}\}$$

3. In example (7),
$$G_0 = \{I, \begin{pmatrix} 1 & 0 \\ 0 & -1 \end{pmatrix}\}.$$

The point group G_0 operates on the lattice T. More precisely, suppose $A \in G_0$ and $t \in T$.

Then $(A, a) \in G$ for some a and

$$(A,a)(I,t)(A,a)^{-1} = (A,At + a)(A^{-1}, -Aa^{-1})$$
$$= (I, -a + At + a)$$
$$= (I,At).$$

So, the translation (I,At) is in G. Since the only translations in G are by vectors from the lattice T, it follows that

$$At \in T.$$

So A is a linear transformation which takes the lattice to itself. In particular, if t_1, t_2 is a lattice basis of T then the matrix of A with respect to t_1, t_2 must be a matrix with integer entries. This is a fact we will use often.

Now, the point group G_0 need not be a subgroup of G --i.e., if $A \in G_0$, then it need not be the case that $(A,0) \in G$ (cf. Example (8)). But the point group is a factor group of G:

Consider the mapping $G \to G_0$
by sending $(A,a) \to A$.

The relations (5) say that this is a homomophism of groups onto G_0. The

82

kernel of this mapping is the set of all elements of the form (I,a) in G
-- that is, the kernel is T. The Fundamental Theorem of group homomophisms then gives

Proposition:

Let G_0 be the point group of G.

Then
$$^G/_T \simeq G_0.$$

Remarks:

1. This shows that the translation subgroup T is a normal subgroup
    of G.

2. Distinct elements of the point group determine distinct cosets
    of T in G.

    Since
    $$(I,t)(A,a) = (A,a + t),$$

    the coset $T(A,a) = \{(A,a + t) | t \varepsilon T\}$.

    Put another way, if $(A,a) \varepsilon$ G,
    then all elements in G that have A for their linear part are of
    the form (A,a + t) for some t.

More can be said:

Proposition:

The point group G_0 of a space group must be a <u>finite</u> group.

Proof:

Consider a circle about the origin containing a lattice basis t_1, t_2
of T. There are only finitely many lattice points inside this

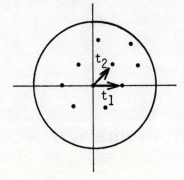

circle -- say n (Note: $n \geq 4$). Since
every $A \varepsilon G_0$ is a distance preserving
linear transformation, every such A
must permute the lattice points
contained in this circle. If the
identity permutation on these n points
results from some A, then A must have
been the identity. This is so, since
such an A would have $At_1 = t_1$ and

83

At $_2$ = t_2 -- and any linear transformation that pointwise fixes two linearly independent elements is automatically the identity. Thus, G_0 is in 1 - 1 correspondence with a subgroup of the permutation group on these n points. The full permutation group is finite and has order n!. In particular, G_0 is finite.

The next question to ask is: What sort of finite groups can occur as point groups?

O(2) and the Crystallographic Restriction

(Consult the Appendix for information on orthogonal matrices)

Here the fact that elements of the point group are distance preserving comes to the fore. The matrix O of a distance preserving linear transformation $A : R^2 \rightarrow R^2$ with respect to the usual basis (1,0) and (0,1) of R^2 is an orthogonal matrix:

$$OO^T = I.$$

The set of all 2 x 2 orthogonal matrices forms a group, designated O(2). If we associate to each A in the point group the orthogonal matrix representing it, we can think of the point group as a subgroup of the orthogonal group O(2). The question we then pose is: What do finite subgroups of O(2) look like?

We know that elements of O(2) are of two types:

rotations through an angle θ,

$R_\theta = \begin{pmatrix} \cos\theta & -\sin\theta \\ \sin\theta & \cos\theta \end{pmatrix}$, which have determimant 1;

reflections through the line making an angle of $\theta/2$ with the positive x axis,

$$F_\theta = \begin{pmatrix} \cos\theta & \sin\theta \\ \sin\theta & -\cos\theta \end{pmatrix}, \text{ which}$$

have determinant -1.

Note that $F_\theta^2 = I.$

We can use this to prove:

Proposition: The finite subgroups of O(2) are either cyclic or dihedral.

84

<u>Proof:</u>   Let G $\neq$ {1} be a finite subgroup of O(2). By the above,
if X ϵ G then det X $=$ $\pm$ 1.

<u>Case 1:</u> <u>All Elements of G have determinant 1.</u>

Then all elements of G are rotations, and among these finitely many
rotations R_Θ there is one with <u>least</u> positive Θ -- say R_{Θ^*}. We claim
that G is cyclic, generated by R_{Θ^*}. Let $R_\Theta \epsilon$ G, $\Theta \neq 0$.
Then Θ is caught between two successive integral multiples of Θ^*:

$$n\Theta^* \leq \Theta < (n + 1)\Theta^*, \text{ for some integer } n \geq 1.$$

Now,

$$R_\Theta R_{\Theta^*}^{-n} \text{ is an element of G.}$$

But then

$$R_\Theta R_{\Theta^*}^{-n} \quad = \quad R_\Theta R_{-n\Theta^*}$$
$$= \quad R_{\Theta - n\Theta^*} \text{ is a rotation}$$

through a positive angle less than Θ^*, <u>unless</u> $\Theta = n\Theta^*$.
So,

$$R_\Theta = R_{n\Theta^*} = R_{\Theta^*}^n .$$

<u>I.e.</u>, every element of G is a power of the rotation R_{Θ^*} and hence G is a
finite cyclic group.

<u>Case 2:</u> <u>Not all elements of G have determinant 1.</u>

Then G contains a reflection, say F.
Consider the mapping G $\rightarrow$ {1, -1} by X $\rightarrow$ detX. This mapping is then a
group homomophism onto {1, -1}. Let H be the kernel of this mapping.
The Fundamental Theorem of Group Homomophisms says that H is a subgroup
of G of index 2. Since the reflection F is not in H, it follows that
the cosets of H in G are precisely H and HF.
Now, H is a finite subgroup of O(2) all of whose elements have det 1.
Thus, by Case 1, H must be cyclic and generated by a rotation, say R.
Suppose R has order n. We have then shown that the list of 2n distinct
elements

I R, R^2, ..., R^{n-1} F, RF, ..., $R^{n-1}F$ accounts for the elements of G.

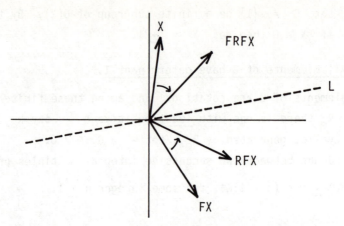

F is a reflection through the
line L. R is a rotation through θ.

A pictorial argument can be used to show that $FRF = R^{-1}$.
Thus,

$$G = \{R^i F^j \mid 0 \le i \le n-1,\ 0 \le j \le 1\}$$

with defining relations

$$R^n = I,\ F^2 = I,\ FRF = R^{-1} = R^{n-1}.$$

Such a group is called a dihedral group.

We will use the notation C_n for a cyclic group of order n and D_n for a
dihedral group of order 2n. A subgroup of O(2) isomorphic to C_n must
contain a rotation through $\frac{2\pi}{n}$ (Exercise 11). The group D_n can be
thought of as the group of symmetries of a regular n-gon: R can be
taken as a rotation through $\frac{2\pi}{n}$ and F as a reflection through an axis of
the n-gon.

So, any point group is isomorphic to C_n or D_n, for some n. Even better,
we can limit the values of n.

Proposition: (Crystallographic Restriction)

Let R be a rotation in a point group through an
angle $\frac{2\pi}{n}$. Then n is 1, 2, 3, 4 or 6.

Proof:

Thinking of R as an element of O(2), its matrix is of the form

$$\begin{pmatrix} \cos\theta & -\sin\theta \\ \sin\theta & \cos\theta \end{pmatrix} \text{ for } \theta = \frac{2\pi}{n}.$$

86

The trace of this matrix is 2 cos Θ. On the other hand, the matrix of R with respect to a lattice basis has integer entries -- and hence an integer trace. But matrices of the same linear transformation with respect to different bases are similar, and similar matrices have the same trace. The upshot is that 2 cos Θ must be an integer. So,

$$\cos\Theta = 0, \pm1, \pm\tfrac{1}{2}.$$

These correspond to the values n = 1, 2, 3, 4 or 6.

Since C_n and D_n both contain rotations through $\dfrac{2\pi}{n}$, the implication for point groups is:

  any point group must be isomorphic to one of the ten groups in the list

(9)
$$C_1, C_2, C_3, C_4, C_6$$
$$D_1, D_2, D_3, D_4, D_6.$$

Crystallographers speak of 10 <u>crystal classes</u> in the plane -- refering to the above 10 possibilities for point groups. No two of the groups in the above list are isomorphic -- with one exception: D_1 is isomorphic to C_2. It will shortly become clearer why we choose to list these two individually.

Having narrowed the range for point groups, the hunt for all possible space groups is well on its way.

Concluding Remarks

Had we pursued our discussion in three dimensions instead of two, the role of O(2) would have been played by the three dimensional orthogonal group O(3). An almost identical argument would have shown that the point group is a finite subgroup of O(3). The difference is that O(3) allows several more types of finite subgroups, yielding 32 crystal classes in three dimensions instead of the ten described in two dimensions.

References pertaining to group theory and crystallography are contained at the end of the next chapter.

1. 1)   Show that $t_1 = \binom{3}{4}$, $t_2 = \binom{5}{7}$

   and $s_1 = \binom{1}{1}$, $s_2 = \binom{1}{0}$

   are both bases for the same lattice in R^2.

   2)   Find a unimodular matrix U with the property that $U \binom{m}{m}$ gives the coordinates of the lattice vector $t = mt_1 + nt_2$ with respect to s_1, s_2.

2.   Verify that E(2) is a group.

3.   Show that $\{T_a | a \in R^2\}$ forms an abelian normal subgroup of E(2).

4.   Is E(2) itself a space group?

5.   Verify the relations in (5).

6.   Show that the sets in examples (6) - (8) are indeed groups.

7.   Find the point groups of the space groups in the previous exercise.

8.   Show that with respect to a lattice basis the matrix of an element in the point group is a unimodular matrix.

9.   How many 2 x 2 matrices are both orthogonal and unimodular?

10.   Explain why C_6 and D_3 are not isomorphic.

11.   Prove that a subroup of O(2) isomorphic to C_n must contain a rotation through $\frac{2\pi}{n}$.

12.   Let T: $R^2 \to R^2$ be a symmetry that satisfies $T(0) = 0$.

   a)   Show that T preserves collinearity: if x, y, z $\in R^2$ are collinear, then T(x), T(y), T(z) are collinear.

   b)   Conclude that
   $$T(\gamma x + (1 - \gamma)y) = \gamma T(x) + (1 - \gamma) T(y),$$
   $\forall\ \gamma \in R$, x, y $\in R^2$.

   c)   Hence, $T(\gamma x) = \gamma T(x)$
   and
   $$T(x + y) = T(x) + T(y).$$
   i.e.,
   T is a linear transformation.

88

13.   A group G is said to be a <u>semidirect product</u> of A by B if G
      contains subgroups A and B satisfying:
      1)   A is normal in G
      2)   AB = G
      3)   A ∩ B = {1}

      Let $T = \{T_a \mid a \in R^2\}$ and A = set of all distance preserving
      linear transformations in R^2. Show that E(2) is a semidirect
      product of T by A.

14.   Let G be the space group in (6). Show that G is a semidirect
      product of its translation subgroup by a subgroup isomorphic to
      its point group.

6 Appendix: Orthogonal Matrices in Two Dimensions

<u>Definition</u>: A 2 x 2 real matrix 0 is <u>orthogonal</u> if

$$00^T = 0^T0 = I \quad \text{(i.e. if } 0^T = 0^{-1}\text{).}$$

<u>Examples and Remarks</u>:

1. $\begin{pmatrix} 1 & 0 \\ 0 & 1 \end{pmatrix}$, $\begin{pmatrix} 0 & 1 \\ -1 & 0 \end{pmatrix}$, $\begin{pmatrix} \frac{1}{2} & \frac{-\sqrt{3}}{2} \\ \frac{\sqrt{3}}{2} & \frac{1}{2} \end{pmatrix}$ are

   orthogonal matrices; $\begin{pmatrix} 1 & 0 \\ 1 & 1 \end{pmatrix}$ is not.

2. Let (x, y) denote the 'usual' scalar product of vectors in R^2:

   if $x = \begin{pmatrix} x_1 \\ x_2 \end{pmatrix}$, $y = \begin{pmatrix} y_1 \\ y_2 \end{pmatrix}$,

   then

   $$(x, y) = x_1 y_1 + x_2 y_2 .$$

   Note that if we write x and y as column vectors, then $(x,y) = x^T y.$

3. If $||x|| = x_1^2 + x_2^2$ denotes the Euclidean length of the vector x in R^2, then

   (1) $(x, x) = ||x||^2 .$

4. Two vectors x, y are <u>perpendicular</u> (or orthogonal)
   $\longleftrightarrow (x, y) = 0.$

5. Phrased another way, a 2 x 2 matrix is orthogonal $\longleftrightarrow$ its columns have length 1 and are perpendicular to each other.

6. If 0 is orthogonal, then

   $$\det(00^T) = \det I = 1.$$

But, $\det(OO^T) = \det O \, \det O^T = (\det O)^2,$

using the fact that $\det A = \det A^T$ for any matrix A.

So, $(\det O)^2 = 1$ and thus $\det O = \pm 1$. I.e., any orthogonal matrix has determinant 1 or -1.

7. The set of all 2 × 2 orthogonal matrices forms a group, called the orthogonal group and denoted O(2).

Describing all 2 × 2 orthogonal matrices is not difficult. Let $O \in O(2)$. Since the first column of O has length 1, it must be of the form

$$\begin{pmatrix} \cos\theta \\ \sin\theta \end{pmatrix}$$

for some θ, $0 \leq \theta < 2\pi$ -- i.e. a vector on the unit circle. There are then only two possibilities for the second column, since it must be a vector on the unit circle perpendicular to the first column.

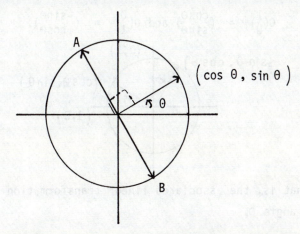

These two possibilities (A and B in the figure) are easily computed to be

$$\begin{pmatrix} -\sin\theta \\ \cos\theta \end{pmatrix} \quad \text{or} \quad \begin{pmatrix} \sin\theta \\ -\cos\theta \end{pmatrix} .$$

Thus, any orthogonal matrix is of the form

$$\begin{pmatrix} \cos\theta & -\sin\theta \\ \sin\theta & \cos\theta \end{pmatrix} \quad \text{for some } \theta, \qquad \text{(Type 1)}$$

or

$$\begin{pmatrix} \cos\theta & \sin\theta \\ \sin\theta & -\cos\theta \end{pmatrix} \quad \text{for some } \theta, \qquad \text{(Type 2)}.$$

91

The choice $\theta = \frac{\pi}{2}$ in type 2 gives the element $\begin{pmatrix} 0 & 1 \\ 1 & 0 \end{pmatrix}$ of 0(2).

The determinant of a Type 1 matrix is $\cos^2\theta + \sin^2\theta$, namely 1. Type 2 matrices have determinant -1.

Geometric Interpretations

Let $0 \in 0(2)$. For $x \in R^2$ (regarded as a column vector), the mapping

$$x \longmapsto 0x$$

is a linear transformation. Such a linear transformation will be uniquely determined by its effect on the basis $\begin{pmatrix} 1 \\ 0 \end{pmatrix}$, $\begin{pmatrix} 0 \\ 1 \end{pmatrix}$ of R^2.

1. If 0 is of Type 1, then

$$0\begin{pmatrix} 1 \\ 0 \end{pmatrix} = \begin{pmatrix} \cos\theta \\ \sin\theta \end{pmatrix} \text{ and } 0\begin{pmatrix} 0 \\ 1 \end{pmatrix} = \begin{pmatrix} -\sin\theta \\ \cos\theta \end{pmatrix}.$$

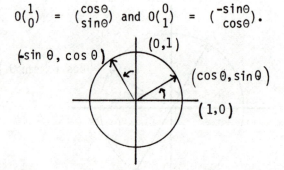

That is, the associated linear transformation is a __rotation__ through an angle θ.

2. A Type 2 matrix 0 can be written in the form

$$\begin{pmatrix} \cos\theta & \sin\theta \\ \sin\theta & -\cos\theta \end{pmatrix} = \begin{pmatrix} \cos\theta & -\sin\theta \\ \sin\theta & \cos\theta \end{pmatrix} \begin{pmatrix} 1 & 0 \\ 0 & -1 \end{pmatrix}$$

The matrix $\begin{pmatrix} 1 & 0 \\ 0 & -1 \end{pmatrix}$ describes a reflection through the x-axis. Thus, a type 2 matrix is a reflection through the x-axis followed by a rotation through the angle θ. Performing these two operations is the same as simply reflecting through the line L making an angle of $\theta/2$ with the positive x-axis:

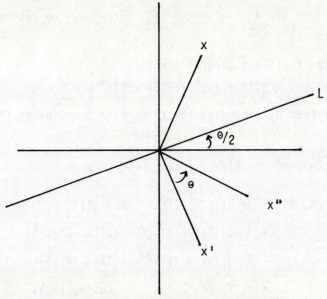

x' is the reflection of x through the horizontal axis.

   x'' is the reflection of x through L.

Type 2 matrices can then be described as reflections.

To summarize: Let $0 \in 0(2)$.

   1.   If $\det 0 = 1$, then 0 is a rotation.

   2.   If $\det 0 = -1$, then 0 is a reflection.

Connection with distance-preserving linear transformations:

   Let $T: R^2 \rightarrow R^2$ be a distance preserving linear
transformation -- i.e.

$$||Tx|| = ||x|| \quad \forall x \in R^2.$$

Let A be the matrix of T with respect to the 'usual' basis $(1, 0)$,
$(0, 1)$ of R^2. So,

$$Tx = Ax \quad \forall x \in R^2.$$

We ask, What sort of matrix is A?

Let $x = \binom{1}{0}$.

Then,

$$||A\binom{1}{0}|| = ||T\binom{1}{0}|| = ||\binom{1}{0}|| = 1.$$

93

So, the first column of A has length 1. Likewise, the second column of A has length 1.

We make use of the following general identity

(2) $(X, Y) = \tfrac{1}{2}\left[||X + Y||^2 + ||X - Y||^2\right]$

which can be verified by expanding the right-hand side, using (1).

Let

$$X = \begin{pmatrix}1\\0\end{pmatrix} \quad Y = \begin{pmatrix}0\\1\end{pmatrix}.$$

then

$$\begin{aligned}
0 = (X, Y) &= \tfrac{1}{2}\left[||X + Y||^2 + ||X - Y||^2\right]\\
&= \tfrac{1}{2}\left[||T(X + Y)||^2 + ||T(X - Y)||^2\right]\\
&= \tfrac{1}{2}\left[||TX + TY||^2 + ||TX - TY||^2\right]\\
&= (TX, TY) \qquad \text{(using (2))}.\\
&= (AX, AY)
\end{aligned}$$

So, $(A\begin{pmatrix}1\\0\end{pmatrix}, A\begin{pmatrix}0\\1\end{pmatrix}) = 0$.

That is, the columns of A are perpendicular. A is thus orthogonal.

We have shown: the matrix of a distance preserving linear transformantion with respect to (1, 0), (0, 1) is orthogonal.

Exercises

1. Verify that 0(2) is a group.

2. Suppose 0 is an orthogonal matrix. Show that
   $(0x, 0y) = (x, y), \ \forall \ x, y \ \varepsilon \ R^2.$

3. Show that the matrices in 0(2) of determinant 1 form a normal subgroup of 0(2), (called the special orthogonal group SO(2)).

4. Find an infinite abelian subgroup of 0(2).

5. Let $\quad 0 = \begin{pmatrix}\cos\theta & \sin\theta\\ \sin\theta & -\cos\theta\end{pmatrix}, \quad x = \begin{pmatrix}\cos\theta/2\\ \sin\theta/2\end{pmatrix}$

   Show that 0x = x. (So that 0 fixes the line L determined by x).

6. Verify the identity (2).

7. Show that the product of two reflections in 0(2) is a rotation.

94

7   Crystallographic Groups in the Plane II

In this chapter we continue with our goal of detailing all possible two dimensional space groups G. We have already established that such a G has a translation subgroup T and associated point group G_0 satisfying $G/T \simeq G_0$.

Intuitively, we will try to recapture G from knowledge of G_0 and T. G_0 is fairly well in hand since G_0 must be isomorphic to one of the ten groups C_1, C_2, C_3, C_4, C_6, D_1, D_2, D_3, D_4, or D_6. Recall also that, if we fix a lattice basis of T, then any element of the point group G_0 can be viewed as a matrix with integer coefficients. We will show that T always possesses a lattice basis for which this matrix representation of G_0 is relatively simple. But first we must agree on what it means to say that two space groups are 'the same'.

Equivalence of Space Groups

We attempt to find all space groups by asking how many space groups are there that admit a given point group. The number of space groups is infinite, but intuitively many of them describe the same types of symmetries -- their differences are non-essential.

For example, take the lattice T with basis (1,0) and (0,1) and the space group G = T consisting purely of the translations allowed by T. Taking (1,0) and (0,2) as a lattice basis, we get another lattice T' and space group G' = T'.

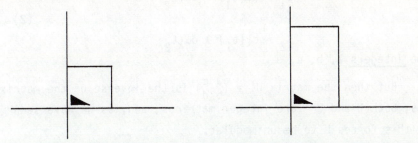

Both G and G' are groups consisting solely of combinations of translations in two independent directions. Pictorially, it is just the shape of a basic parrallelogram that has changed -- the types of symmetries the crystal allows remain the same.

Loosely, two space groups are the same if a possible change of frame of reference results in the same types of symmetries. We express this by

<u>Def.</u>   Let G and G' be space groups with respective translation
          subgroups (lattices) T and T'.
(1)
          G and G' are <u>equivalent</u> if there exists a group isomorphism
               ϕ: G $\dashrightarrow$ G' with $\phi(T) = T'$.

<u>Example</u>:

In the above, the mapping
     $(I, m(1, 0) + n(0,1)) \dashrightarrow (I, m(1,0) + n(0, 2))$
shows that G and G' are equivalent. As a matter of fact, any two space
groups consisting solely of translations are equivalent.

          This definition of equivalence has some immediate
consequences.
          Suppose G and G' are equivalent space groups with $\phi(T) = T'$.
Let t_1, t_2 be a lattice basis of T and
     s_1, s_2 be a lattice basis of T'.
Then we can write
     $\phi(t_1) = \alpha s_1 + \beta s_2$ and $\phi(t_2) = \gamma s_1 + \delta s_2$
for some integers α, β, γ, δ.
The matrix $U = \begin{pmatrix} \alpha & \gamma \\ \beta & \delta \end{pmatrix}$ has the property that if
$t = mt_1 + nt_2$, then $U\begin{pmatrix} m \\ n \end{pmatrix}$ is the column vector giving the coefficients of
$\phi(t)$ with respect to s_1 and s_2.
          Since $\phi(T) = T'$, it follows that $\phi(t_1)$, $\phi(t_2)$ also form a
lattice basis of T'. So, we can write
               $s_1 = a\phi(t_1) + b\phi(t_2)$
and   (2)
               $s_2 = c\phi(t_1) + d\phi(t_2)$
for some <u>integers</u> a, b, c, d.

          But then the matrix $U' = \begin{pmatrix} a & c \\ b & d \end{pmatrix}$ is the inverse of the matrix
U. Thus, both U and U^{-1} are integer matrices, and, as we have seen
before, this forces U to be unimodular.

          We have shown that
               $\phi(t) = Ut, t \in T,$
where U is a unimodular matrix and $t = \begin{pmatrix} m \\ n \end{pmatrix}$ as above.

          The matrix U also provides a relation between the point groups
of G and G'. Think of an element A in the point group of G as a square
matrix with integer coefficients (represent it in the basis t_1, t_2).

Suppose (A, a) in G is sent by ϕ to (A', a') in G'. Likewise, w.r.t. s_1, s_2, A' is an integer matrix. Let $t \in T$.

Then
$$(A, a) (I, t) (A, a)^{-1}$$
must be sent to
$$(A', a') (I, Ut) (A', a')^{-1} \text{ by } \phi.$$
Evaluating both expressions, we get
$$\phi((I, At)) = (I, A'Ut)$$
i.e.,
$$(I, UAt) = (I, A'Ut)$$
So,
$$UAt = A'Ut, \quad \forall t \in T$$
This forces
$$UA = A'U$$
or
$$A' = UAU^{-1}.$$
This says that every element in the point group of G' must be conjugate to an element of the point group of G by the unimodular matrix U. We have

<u>Proposition:</u>   Let G_0 and G_0' be the point groups of G and G'. If G and G' are equivalent, then there exists a unimodular matrix U with $G_0' = UG_0U^{-1}$. (I.e., the point groups are <u>unimodularly equivalent.</u>)

<u>Remarks:</u>

1. In speaking of point groups, we will often use the shorter phrase <u>equivalent</u> instead of unimodularly equivalent. Saying two point groups are equivalent is the same as saying that their respective matrices w.r.t. lattice bases are similar via a unimodular matrix. Note that matrix U depends on the choice of lattice bases in T and T'.

2. Suppose G and G' have respective point groups
$$G_0 = \{I, \begin{pmatrix} -1 & 0 \\ 0 & -1 \end{pmatrix}\} \text{ and } G_0' = \{I, \begin{pmatrix} 0 & 1 \\ 1 & 0 \end{pmatrix}\}.$$
G_0' is <u>not</u> equivalent to G_0, since if
$$\begin{pmatrix} 0 & 1 \\ 1 & 0 \end{pmatrix} = U\begin{pmatrix} -1 & 0 \\ 0 & -1 \end{pmatrix}U^{-1}$$
for some U, it would follow that det $\begin{pmatrix} 0 & 1 \\ 1 & 0 \end{pmatrix} = \begin{pmatrix} -1 & 0 \\ 0 & -1 \end{pmatrix}$, which is not true. (Recall that similar matrices have equal determinants and

traces). Note, though, that G_0 and G_0' are isomorphic as abstract groups -- they are both cyclic of order 2.

3. The above shows that equivalence of point groups is a stricter classification scheme than isomorphism. In our program of classifying space groups we must then take a closer look at the 10 crystal classes of point groups and see how they break up under unimodular equivalence. Here's another way of stating it: any point group G_0 fixes a lattice T. If $t \in T$ and $A \in T$, then $At \in T$. With respect to a lattice basis of T the matrix of A has integer entries. G_0 can thus be thought of as a group of matrices with integer coefficients. Two point groups G_0 and G_0' are equivalent if they are conjugate as subgroups of the group of all 2 x 2 unimodular matrices.

4. Any space group G is a subgroup of E(2). Conjugation by an element $g = (X, s)$ of E(2) will always produce a group gGg^{-1} which is isomorphic as an abstract group to G.

We note that if $X = U$, a <u>unimodular</u> matrix, then the groups G and $G' = (U, s)G(U,s)^{-1}$ are equivalent space groups. This is so, since
$$(U, s)(I, t)(U,s)^{-1} = (I, Ut).$$
Thus, the lattice T' of G' satisfies $T' = UT$.

For example, if T is the lattice with basis (1, 0) and (0, 1) and G is the space group consisting of T and all elements of the form
$$\left(\begin{pmatrix} -1 & 0 \\ 0 & -1 \end{pmatrix}, t + \begin{pmatrix} 1/2 \\ 0 \end{pmatrix}\right), t \in T,$$
then taking $\qquad\qquad\qquad\qquad\qquad\qquad\qquad\qquad\qquad\qquad$ (3)
$$U = I, s = \begin{pmatrix} -1/4 \\ 0 \end{pmatrix},$$
we have that
$$(I, s)G(I, s)^{-1}$$
is the space group consisting of T and all elements of the form
$$\left(\begin{pmatrix} -1 & 0 \\ 0 & -1 \end{pmatrix}, t\right), t \in T.$$
These two space groups are thus equivalent.

The mapping $x \longrightarrow (I, s) x (I, s)^{-1}$ affects a shift of the coordinate system by the vector s. Geometrically, this example says that the patterns

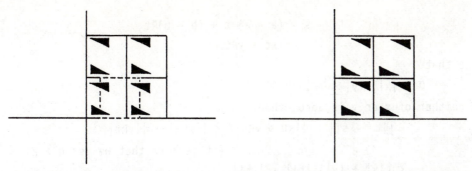

have the same symmetry, and that this can be seen more clearly by using
the outlined square in the first to replicate the pattern.

Next we examine the types of lattices that can occur and the
way in which point groups can act on them.

Types of Lattices

Proposition   Let R be a rotation through $\frac{2\pi}{n}$ in the point group G_0,
    (4)   t a non-zero lattice vector of minimal length.
                    If $n > 2$, then t, Rt form a lattice basis.

Proof:

Let t_1, t_2 be a lattice basis. If t_1 and t_2 were each integral
linear combinations of t and Rt, then every lattice vector could be
expressed as an integral linear combinatiion of t and Rt and we would be
done. We show that t_1 can indeed be written this way. The argument for
t_2 is identical. Now, we can write

$$t = \alpha t_1 + \beta t_2$$
$$Rt = \gamma t_1 + \delta t_2,$$

for some __integers__ α, β, γ, δ.
Solving the above equations for t_1, we can write

$$t_1 = at + bRt,$$

where a and b are __rational__.

We show that a and b are actually integers. Let $\overline{a}$ denote the __integer
nearest__ to a:

$$\overline{\frac{2}{3}} = 1, \quad \overline{\frac{1}{2}} = 1, \quad \overline{\frac{1}{3}} = 0.$$

Then,

$$s = \overline{a}\, t + \overline{b}\, Rt \text{ is a lattice vector.}$$

and so is

99

$$t_1 - s = (a - \bar{a})\, t + (b - \bar{b})Rt$$
$$= xt + yRt.$$

Note that
$$0 \leq |x|, \ |y| \leq \tfrac{1}{2}.$$

If neither of x or y is zero, then
$$||t_1 - s|| = ||xt + yRt|| < ||xt|| + ||yRt||$$

(it is here that we use n > 2)

$$\leq (|x| + |y|)||t|| \leq ||t||.$$

So $t_1 - s$ is a non-zero lattice vector of length less than that of t -- a contradiction. Thus, one of x or y is zero, say x. It follows easily that y is also zero. That is, a and b must be integers.

We now examine what can be said if the point group G_0 contains a reflection F. Choose axes so that the x-axis is the axis of reflection.

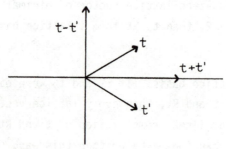

Let t be a lattice vector not on either of the co-ordinate axes and t' its image under F.

Then t + t' sits on the x-axis, while t - t' is on the y-axis.

Let t_1 and t_2 denote lattice vectors of <u>minimum</u> length on the x and y axes respectively

Then
$$t + t' = mt_1$$
$$t - t' = nt_2.$$

for some <u>integers</u> m and n.

(Otherwise, e.g., if $t + t' = 1.7t_1$, then $t + t' - t_1 = .7t_1$ would be a horizontal lattice vector shorter than t_1).

So,
$$t = \tfrac{m}{2}t_1 + \tfrac{n}{2}t_2. \tag{5}$$

100

It cannot happen that of m and n, one is even while the other is odd (cf. Exercise 5).

If m and n are always both even, then t_1, t_2 form a lattice basis.

Otherwise, nothing that

$$\frac{m}{2}t_1 + \frac{n}{2}t_2 = (\frac{m+n}{2}) (\frac{t_1}{2} + \frac{t_2}{2}) + (\frac{m-n}{2}) (\frac{t_1}{2} - \frac{t_2}{2})$$

we get that

$$\frac{t_1}{2} + \frac{t_2}{2} , \frac{t_1}{2} - \frac{t_2}{2}$$

form a lattice basis.

<u>To Summarize</u>: If G_0 contains a reflection F, then either

1) 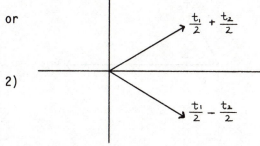 The lattice is <u>rectangular</u> with the side of a rectangle as an axis of reflection.

or

2) 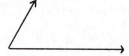 The lattice is <u>rhombic</u> with the diagonal an axis of reflection.

Keeping in mind the allowable types of point groups, we then have the following types of lattices that can occur in space groups.

		Point Group
1. General		C_1, C_2
2. Rectangular	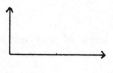	D_1, D_2
3. Rhombic		D_1, D_2

4. Square C_4, D_4

5. Hexagonal 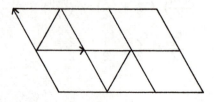 C_3, D_3
C_6, D_6

Point Groups Admitted by Lattices

Any point group acts on a lattice which must be one of the five previously determined types. We can thus give a matrix representation of the point group relative to a basis t_1, t_2 of the lattice.

1. <u>General</u>

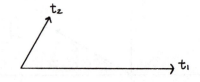

1) $C_1 = \{I\} = \{\begin{pmatrix} 1 & 0 \\ 0 & 1 \end{pmatrix}\}$

2) $C_2 = \{I, \begin{pmatrix} -1 & 0 \\ 0 & -1 \end{pmatrix}\}$

2. <u>Rectangular</u>

1) Only one side is an axis of reflection:

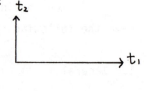

$D_{1,r} = \{I, \begin{pmatrix} 1 & 0 \\ 0 & -1 \end{pmatrix}\}$

2) Both sides are axes of reflection:

$D_{2,r} = \{I, \begin{pmatrix} 1 & 0 \\ 0 & -1 \end{pmatrix}, \begin{pmatrix} -1 & 0 \\ 0 & 1 \end{pmatrix}, \begin{pmatrix} -1 & 0 \\ 0 & -1 \end{pmatrix}\}$

3. <u>Rhombic</u>

1) Only one diagonal is an axis of reflection:

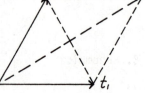

$D_{1,rh} = \{I, \begin{pmatrix} 0 & 1 \\ 1 & 0 \end{pmatrix}\}$

2) Both diagonals are axes of reflection:

$D_{2,rh} = \{I, \begin{pmatrix} 0 & 1 \\ 1 & 0 \end{pmatrix}, \begin{pmatrix} 0 & -1 \\ -1 & 0 \end{pmatrix}, \begin{pmatrix} -1 & 0 \\ 0 & -1 \end{pmatrix}\}$

4. <u>Quadratic</u>

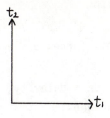

1) Rotations only -- a cyclic group of order 4:
$$C_4 = \{I, \begin{pmatrix} 0 & -1 \\ 1 & 0 \end{pmatrix}, \begin{pmatrix} -1 & 0 \\ 0 & -1 \end{pmatrix}, \begin{pmatrix} 0 & 1 \\ -1 & 0 \end{pmatrix}\}$$

2) Allow also reflection in one of the sides:
$$D_4 = C_4 \cup \{\begin{pmatrix} 1 & 0 \\ 0 & -1 \end{pmatrix}, \begin{pmatrix} 0 & -1 \\ -1 & 0 \end{pmatrix}, \begin{pmatrix} -1 & 0 \\ 0 & 1 \end{pmatrix}, \begin{pmatrix} 0 & 1 \\ 1 & 0 \end{pmatrix}\}$$

5. <u>Hexagonal</u>

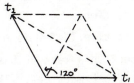

1) <u>Cyclic group of order 3</u>
$$C_3 = \{I, \begin{pmatrix} 0 & -1 \\ 1 & -1 \end{pmatrix}, \begin{pmatrix} -1 & 1 \\ -1 & 0 \end{pmatrix}\}$$

2) <u>Cyclic Group of order 6</u>
$$C_6 = C_3 \cup \{\begin{pmatrix} 1 & -1 \\ 1 & 0 \end{pmatrix}, \begin{pmatrix} -1 & 0 \\ 0 & -1 \end{pmatrix}, \begin{pmatrix} 0 & 1 \\ -1 & 1 \end{pmatrix}\}$$

3) Rotation through 120° plus reflection in the
   a) <u>shorter diagonal</u>

$$D_{3,s} = C_3 \cup \{\begin{pmatrix} 0 & 1 \\ 1 & 0 \end{pmatrix}, \begin{pmatrix} 1 & -1 \\ 0 & -1 \end{pmatrix}, \begin{pmatrix} -1 & 0 \\ -1 & 1 \end{pmatrix}\}$$

   b) <u>longer diagonal</u>

$$D_{3,\ell} = C_3 \cup \{\begin{pmatrix} 0 & -1 \\ -1 & 0 \end{pmatrix}, \begin{pmatrix} -1 & 1 \\ 0 & 1 \end{pmatrix}, \begin{pmatrix} 1 & 0 \\ 1 & -1 \end{pmatrix}\}$$

4) <u>Rotation through 60° plus reflection in one of the diagonals:</u>
$$D_6 = C_6 \cup \{\begin{pmatrix} 0 & 1 \\ 1 & 0 \end{pmatrix}, \begin{pmatrix} 1 & -1 \\ 0 & -1 \end{pmatrix}, \begin{pmatrix} -1 & 0 \\ -1 & 1 \end{pmatrix}, \begin{pmatrix} 0 & -1 \\ -1 & 0 \end{pmatrix}, \begin{pmatrix} -1 & 1 \\ 0 & 1 \end{pmatrix}, \begin{pmatrix} 1 & 0 \\ 1 & -1 \end{pmatrix}\}$$

It is important to note that <u>no two</u> of the above thirteen matrix groups are unimodularly equivalent. For example, among pairs of groups that are isomorphic as abstract groups, C_2 and $D_{1,r}$ are not equivalent since $\begin{pmatrix} -1 & 0 \\ 0 & -1 \end{pmatrix}$ and $\begin{pmatrix} 1 & 0 \\ 0 & -1 \end{pmatrix}$ have distinct traces (or determinants).

If $D_{1,r}$ and $D_{1,rh}$ were equivalent, then there would exist a unimodular matrix
$$\begin{pmatrix} \alpha & \beta \\ \gamma & \delta \end{pmatrix}$$

with

$$\begin{pmatrix} \alpha & \beta \\ \gamma & \delta \end{pmatrix} \begin{pmatrix} 1 & 0 \\ 0 & -1 \end{pmatrix} = \begin{pmatrix} 0 & 1 \\ 1 & 0 \end{pmatrix} \begin{pmatrix} \alpha & \beta \\ \gamma & \delta \end{pmatrix}$$

This forces $\alpha = \gamma$ and $\beta = -\delta$, so that

$$\det \begin{pmatrix} \alpha & \beta \\ \gamma & \delta \end{pmatrix} = -2\alpha\beta .$$

By unimodularity, $-2\alpha\beta = \pm 1$, which can't happen since α and β are integers.

The Classification of Space Groups

Our hunt for all space groups begins with the point group. Given G_0, we wish to find all space groups G whose point group is G_0. Some observations:

1) The lattice T of G is one of the types listed previously. With respect to a lattice basis of T the elements of G_0 can be written as integer mattices. This basis can be chosen so that G_0 is one of the 13 previously derived matrix groups.

2) We write $G_0 = \{A_1, A_2, \ldots, A_n\}$ where the A_i are matrices.

3) G can also be written in this matrix form. An element of G is then of the form (A_i, a), where a is also expressed relative to the lattice basis.

4) We ask, what are the possibilities for G? I.e., we want all inequivalent space groups G, with $G/T \simeq G_0$.

5) Any such G admits a coset decomposition

$$G = T(A_1, a_1) \cup T(A_2, a_2) \cup \ldots \cup T(A_n, a_n)$$

for some column vectors a_i.

6) The problem then reduces to asking, what choices of a_i are permissible?

7) One choice is always possible: take $a_i = 0$, for all i. This yields the space group

$$G = \{(A, t) \mid A \varepsilon G_0, t \varepsilon T\}$$

In group theoretic language, G is the semi-direct product of G_0 and T; crystallographers refer to such G's as symmorphic space groups. Following Burckhardt [3] we will refer to the choice $a_i = 0$ as the 'trivial solution'.

8) There are thus at least 13 non-equivalent 2-dimensional space groups corresponding to the trivial solution for each class of non-equivalent point groups. We will show that there are only four more types of space groups that can occur.

9) Since $T(A, a) = T(A, a + t)$ for any $t \in T$, we may assume that

$$a_i = \binom{x_i}{y_i} \text{ has } 0 \leq x_i, y_i < 1.$$

(We are using a lattice basis: the t's are all vectors with integer coordinates.)

10) The a_i's must satisfy certain relations:

if $A_i A_j = A_k$ in G_0,

then, since

$$(A_i, a_i) (A_j, a_j) = (A_k, A_i a_j + a_j),$$

5) then forces

$$A_i a_j + a_i = a_k + t,$$

for some $t \in T$.

That is, the two vectors $A_i a_j + a_i$ and a_k must differ by a vector with integer coordinates.

We will indicate this by writing

$$A_i a_j + a_i \equiv a_k \pmod{1},$$

if $\hspace{10cm}$ (*)

$$A_i A_j = A_k.$$

11) $(U, s) G (U, s)^{-1}$ is a space group equivalent to G, if U is a unimodular matrix.

Let

$$G = \bigcup_{i=1}^{n} T (A_i, a_i),$$

and

$$G' = \bigcup_{i=1}^{n} T (A_i, b_i)$$

be space groups with the same point groups (expressed using the same lattice T).

Note that

$$(I, s) (A_i, a_i) (I, s)^{-1}$$
$$= (A_i, a_i + s) (I, -s)$$
$$= (A_i, -A_i s + a_i + s)$$

105

$$= (A_i, a_i - (A_i - I)s).$$

Thus, if there exists an s with

(6) $\qquad a_i - (A_i - I)s \equiv b_i \pmod 1, \forall i,$

then G and G' are equivalent. In particular, if there exists an s with

(7) $\qquad (A_i - I)s \equiv a_i \pmod 1, \forall i$

then G is equivalent to the trivial solution.

We now make use of the above observations to find all space groups corresponding to a given point group.

C_1

We have already seen that there is only one space group, up to equivalence, with point group C_1. Here, $G = T$.

$D_{1,r}$

Here, $G_0 = \{I, A\}$ where $A = \begin{pmatrix} 1 & 0 \\ 0 & -1 \end{pmatrix}$.

Let $(A,a) \varepsilon G$, where $a = \begin{pmatrix} x \\ y \end{pmatrix}$.

Then

$$A - I = \begin{pmatrix} 0 & 0 \\ 0 & -2 \end{pmatrix}$$

and

$$\begin{pmatrix} x \\ y \end{pmatrix} - \begin{pmatrix} 0 & 0 \\ 0 & -2 \end{pmatrix} \begin{pmatrix} 0 \\ -\frac{y}{2} \end{pmatrix} = \begin{pmatrix} x \\ 0 \end{pmatrix}.$$

So, from (6), with $s = (-\frac{y}{2})$, we may assume that $a = \begin{pmatrix} x \\ 0 \end{pmatrix}$.
The fundamental relation (*) becomes

$$Aa + A \equiv 0 \pmod 1.$$

So, $2x \equiv 0 \pmod 1$, and since $0 \le x < 1$, we have

$$x = 0 \text{ or } x = 1/2.$$

This gives rise to the two possibilities

1) $G_1 = TUT (A,0)$ (the trivial solution)

and

2) $G_2 = TUT (A, \begin{pmatrix} 1/2 \\ 0 \end{pmatrix})$.

Suppose G_1 and G_2 were equivalent space groups with $\phi: G_2 \to G_1$ an isomorphism and $\phi(T) = UT$, for some unimodular U.

Let $\phi(A, \begin{pmatrix} 1/2 \\ 0 \end{pmatrix}) = (A, \begin{pmatrix} u \\ v \end{pmatrix})$, where u and v are integers.

Then since

$$(A, \begin{pmatrix} 1/2 \\ 0 \end{pmatrix}))^2 = (I, \begin{pmatrix} 1 \\ 0 \end{pmatrix})$$

it would follow that

$$\phi(I, \begin{pmatrix} 1 \\ 0 \end{pmatrix})) = (A, \begin{pmatrix} u \\ v \end{pmatrix}))^2 = (I, \begin{pmatrix} 2u \\ 0 \end{pmatrix})).$$

So,

$$U\begin{pmatrix} 1 \\ 0 \end{pmatrix} = \begin{pmatrix} 2u \\ 0 \end{pmatrix}.$$

This says that the first column of U is $\begin{pmatrix} 2u \\ 0 \end{pmatrix}$.

Since U has integer entries, this implies $|\det U| \geq 2$.

But this contradicts the fact that $\det U = \pm 1$.

So in this case we have <u>two</u> inequivalent space groups G_1 and G_2. Below are fundamental parallelograms for each.

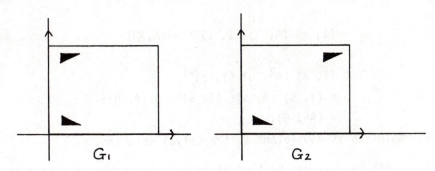

G_1 G_2

$\underline{D_{1, rh}}$

$G_0 = \{I, A\}$ where $A = \begin{pmatrix} 0 & 1 \\ 1 & 0 \end{pmatrix}$.

Letting $a = \begin{pmatrix} x \\ y \end{pmatrix}$, the relation (*) gives $x + y \equiv 0 \pmod 1$.

So, we may take $a = \begin{pmatrix} x \\ -x \end{pmatrix}$.

Noting that

$$(A - I) \begin{pmatrix} 0 \\ x \end{pmatrix} = \begin{pmatrix} x \\ -x \end{pmatrix},$$

using (7) with $s = \begin{pmatrix} 0 \\ x \end{pmatrix}$, we see that any solution is equivalent to the trival solution.

This produces but one group represented by

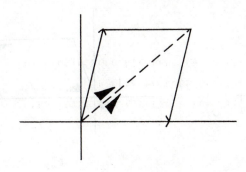

C_2, C_3, C_4, C_6

We argue using the point group C_3. Consider the generator $A = \begin{pmatrix} 0 & -1 \\ 1 & -1 \end{pmatrix}$ of C_3. The $A^3 = I$ and A, being a rotation, fixes no lattice vectors except zero; _i.e._ A - I is an invertible matrix.

Let (I, 0), (A, a), (A^2, b) be a set of coset representatives of T in G. Since $(A, a)^2 = (A^2, Aa + a)$, we may take b = Aa + a.

Letting

$$s = (A - I)^{-1}a$$

we have that

$$(A - I)s = a.$$

Thus,

$$(I, s)\,(A, a)\,(I, s)^{-1} = (A, 0)$$

and

$$(I, s)\,(A^2, b)\,(I, s)^{-1}$$
$$= (I, s)\,(A, a)^2\,(I, s)^{-1} = (A, 0)^2$$
$$= (A^2, 0).$$

Thus any solution is equivalent to the trival solution.

That C_2, C_4, and C_6 also yield only the trival solution can be shown in the same way. The point to note is that each of these cyclic groups is generated by an element A such that A - I is invertible.

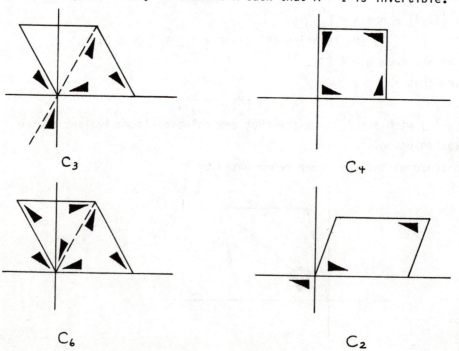

C_3

C_4

C_6

C_2

108

The remaining point groups are dihedral groups which contain cyclic subgroups of index 2.

$D_{3,s}$

Let $A = \begin{pmatrix} 0 & -1 \\ 1 & -1 \end{pmatrix}$, $B = \begin{pmatrix} 0 & 1 \\ 1 & 0 \end{pmatrix}$.

Then, $D_{3,s} = \{I, A, A^2, B, AB, A^2B\}$

Let G be a space group with point group $D_{3,s}$.

Let (A, a_1), (A^2, a_2), (B, b) be elements of G.

The subgroup G' of G generated by T, (A, a_1) and (A^2, a_2) is a space group with point group C_3.

We showed previously that, for some s, (I, s) G' $(I, s)^{-1}$ yields the trivial solution.

Hence, by considering (I, s) G $(I, s)^{-1}$, which is equivalent to G, we may assume that $a_1 = a_2 = 0$.

Since G is generated by T, $(A, 0)$ and (B, b), if we could find a v that simultaneously satisfies

$$(A - I) v \equiv 0 \pmod 1$$

and $\qquad\qquad\qquad\qquad\qquad\qquad\qquad\qquad\qquad\qquad$ (8)

$$(B - I) v \equiv b \pmod 1$$

then (7) would say that G is equivalent to the trivial solution.

We proceed to show such a v exists. Now, the relation $BA = A^2B$ applied to (*) produces

$$A^2b \equiv b \pmod 1$$

So,

$$A^3b \equiv Ab \pmod 1$$

i.e.,

$$b \equiv Ab \pmod 1, \quad \text{since } A^3 = I$$

or

$$(A - I)b \equiv 0 \pmod 1.$$

Since

$$B^2 = I, \text{ (*) yields } Bb + b \equiv 0 \pmod 1$$

Now,

$$B - I = A^2 + A + B,$$

so

$$\begin{aligned}(B - I)b &\equiv (A^2 + A + B)b \pmod 1 \\ &\equiv A^2b + Ab + Bb \pmod 1\end{aligned}$$

109

$$\equiv b + (b + Bb) \quad (\text{mod } 1)$$
$$\equiv b + 0 \quad (\text{mod } 1)$$
$$\equiv b \quad (\text{mod } 1)$$

In other words, the choice $v = b$ provides a solution to (8). There is thus only one space group corresponding to the point group $D_{3,s}$.

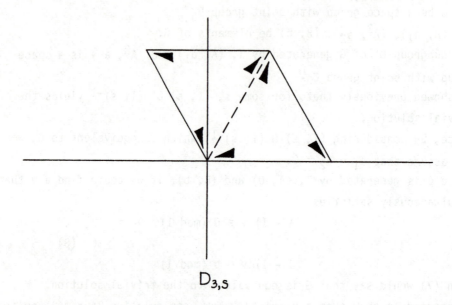

$D_{3,s}$

Similar arguments may be used with $D_{3\ell}$, D_6 and $D_{2,rh}$ to show that each admits only the trivial solution.

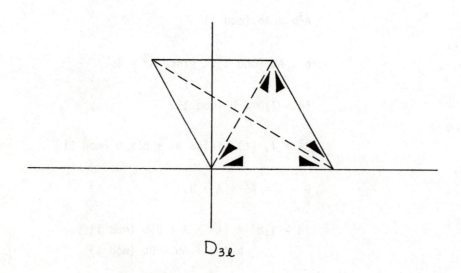

$D_{3\ell}$

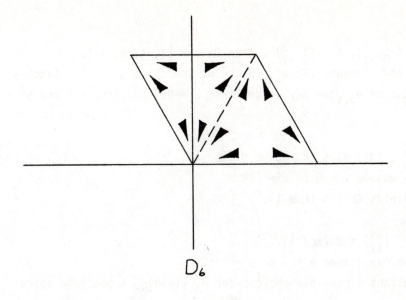

D_6

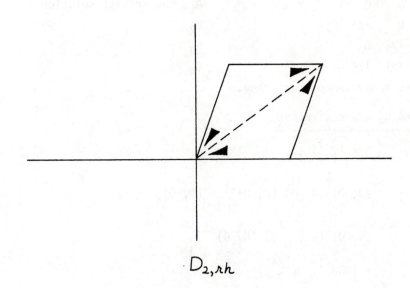

$D_{2,rh}$

There are two point groups remaining.

111

$\underline{D_{2,r}}$

Let $A = \begin{pmatrix} -1 & 0 \\ 0 & -1 \end{pmatrix}$, $B = \begin{pmatrix} 1 & 0 \\ 0 & -1 \end{pmatrix}$. Then $D_{2,r} = \{I, A, B, AB\}$.
Consider the elements (A, a), (B, b) of the space group G. Exactly as
in the case of $D_{3,s}$ we may assume that $a = 0$. Now from (*) and $B^2 = I$,
we get

$$Bb + b \equiv 0 \pmod 1.$$

If $b = \begin{pmatrix} x \\ y \end{pmatrix}$, this says that $2x \equiv 0 \pmod 1$.
So we may assume $x = 0$ or $x = 1/2$.
$AB = BA$ yields $Ab \equiv b \pmod 1$.
Thus,

$$\begin{pmatrix} 2x \\ 2y \end{pmatrix} \equiv 0 \pmod 1.$$

So also we may assume $y = 0$ or $y = 1/2$.
There are thus 4 possible choices for b, yielding 4 possible space
groups.

	x	y		Space Groups	
1)	0	0		G_1, the trivial solution	
2)	0	1/2		G_2	(9)
3)	1/2	0		G_3	
4)	1/2	1/2		G_4	

We examine equivalence among them.

1) $\underline{G_2 \text{ and } G_3 \text{ are equivalent}}$

Let $\qquad U = \begin{pmatrix} 0 & 1 \\ 1 & 0 \end{pmatrix}$.
Then

$$(U, 0)(A, 0)(U, 0)^{-1} = (A, 0),$$

and

$$(U, 0)(B, \begin{pmatrix} 1/2 \\ 0 \end{pmatrix})(U, 0)^{-1}$$

$$= (UBU^{-1}, U\begin{pmatrix} 1/2 \\ 0 \end{pmatrix}))$$

$$= (\begin{pmatrix} -1 & 0 \\ 0 & 1 \end{pmatrix}, \begin{pmatrix} 0 \\ 1/2 \end{pmatrix})) = (AB, \begin{pmatrix} 0 \\ 1/2 \end{pmatrix})$$

and

$$(U, 0)(AB, \begin{pmatrix} 1/2 \\ 0 \end{pmatrix})(U, 0)^{-1}$$

$$= (B, \begin{pmatrix} 0 \\ 1/2 \end{pmatrix}))$$

Thus,

$$UG_3U^{-1} = G_2.$$

2) In G_3, $(B, \binom{1/2}{0}))^2 = (I, \binom{1}{0}))$. Applying exactly the same argument as used in the case $D_{1,r}$ shows that G_3 cannot be equivalent to G_1. Likewise G_4 is not equivalent to G_1.

3) <u>G_4 is not equivalent to G_2</u>

G_2 contains $(A, 0)$ and $(B, \binom{1/2}{0}))$, where

$(A, 0)^2 = (B, \binom{0}{1/2}))^2 = (I, 0)$.

The elements of G_4 are of the form

(I, t), (A, t), $(B, \binom{1/2}{1/2} + t)$, $(AB, \binom{1/2}{1/2} + t)$

as t ranges through T.

Were G_4 and G_2 equivalent, then G_4 would contain two elements which are in distinct cosets of T and whose squares are the identity $(I, 0)$. A check above shows the that G_4 has no two such elements.

The point group $D_{2,r}$ thus produces three inequivalent space groups:

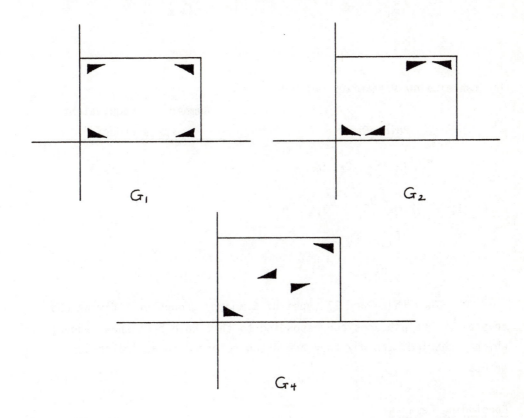

G_1 G_2

G_4

113

<u>D$_4$</u>

The arguments here are similar to those in the above case and are left as an exercise. There are two inequivalent point groups.

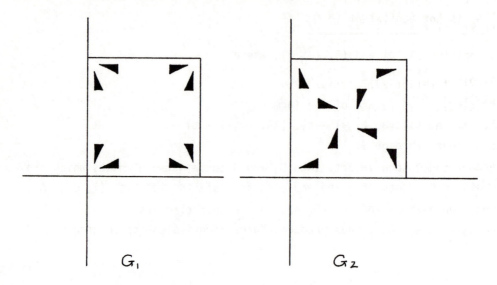

G_1 G_2

The table below summarizes our results:

Point Group	Number of inequivalent space groups
C_1, C_2, C_3, C_4, C_6	1
$D_{1,rh}$, $D_{2,rh}$, $D_{3,s}$, $D_{3,\ell}$, D_6	1
$D_{1,r}$, D_4	2
$D_{2,r}$	3

Thus there are 17 types of symmetry groups of 2 dimensional crystals. The patterns corresponding to them have been shown above, which might indicate why they are often referred to as 'wallpaper groups'.

Concluding Remarks:

In three dimensions there are 32 crystal classes and 219 types

114

of crystallographic space groups (230, if a slightly different definition of equivalence is used). The determination of these groups dates back to the work of Sohncke, Fedorov and Schonflies (1890). Crystallographic groups can be defined in any dimension n, and a famous result due to Bieberbach (1910) states that for any n there are only finitely many space group types. Recent computations (1971) have shown that for n = 4 there are 4783 types [1]. Our classification of space groups for n = 2 is based on the development in [3] which also contains the derivations of the three dimensional groups.

The plane crystallographic groups can be viewed as symmetry groups of periodic designs. A nice commentary on this connection between mathematics and art with references to the work of M.C. Escher is contained in [10]. The article [9] outlines an algorithm that, given a repeating pattern, identifies its symmetry group.

Practical space group determination for real crystals relies on methods of x-ray diffraction. It is interesting to note that the theoretical classification of space groups was accomplished well before x-ray experiments that confirmed the lattice structure of crystals were conducted in 1912 [2].

References

[1] H. Brown, R. Bulow et. al.,   Crystallographic Groups of Four-
                                       Dimensional Space, John Wiley &
                                       Sons, New York, 1978.

[2] M. J. Buerger,   Elementary Crystallography, MIT
                                       Press, Cambridge 1978.

[3] J. J. Burckhardt,   Die Bewegungsgruppen der
                                       Kristallographie, Birkhauser,
                                       Basel, 1957.

[4] D. R. Farkas,   "Crystallographic Groups and
                                       their Mathematics", Rocky
                                       Mountain Journal of Mathematics
                                       Vol. 11, No. 4, 1981, p. 511-551.

[5] P. Le Corbeiller,   "Crystals and the Future of
                                       Physics", Scientific American,
                                       Jan. 1953, p. 50-51.

[6] E. H. Lockwood & R. H. MacMillan, Geometric Symmetry, Cambridge
                                       University Press, 1978

[7] W. Miller, Jr.   Symmetry Groups and their
                                       Applications, Academic Press, New
                                       York, 1972.

[8] G. Polya   "Über die Analogie der
                                       Kristallsymmetrie in der Ebene"
                                       Zeitschrift fur Kristallographie
                                       60 (1924), p. 278-82

[9] B. Rose & R. Stafford   "An Elementary Course in
                                   Mathematical Symmetry" American
                                   Mathematical Monthly, January
                                   1981, p. 59-64.

[10] D. Schattschneider   "The Plan Symmetry Groups: Their
                                   Recognition and Notation"
                                   American Mathematical Monthly,
                                   June-July 1978, p. 439-450

[11] R. L. Schwarzenberger   "The 17 Plane Symmetry Groups"
                                   Mathematical Gazette, Vol 58,
                                   1974, p. 123-131.

[12] H. Weyl,   Symmetry, Princeton University
                                   Press, 1952.

Exercises

1. Show that the matrices U and U' introduced in (2) are inverses of each other.

2. As in Definition (1), let ϕ: G $\rightarrow$ G' give an equivalence of the space groups G and G'. Suppose that t_1, t_2 is a lattice basis of T. Explain why $\phi(t_1)$, $\phi(t_2)$ is a lattice basis of T'.

3. Show that the set 2 x 2 unimodular matrices forms a group.

4. The proof following (4) shows that x = 0. Explain why it follows that also y = 0.

5. In (5), show that it cannot be that m is even, while n is odd.

6. (Refers to the section <u>Point Groups Admitted by Lattices</u>) Verify the matrix representations given for the following point groups with respect to the sketched lattice bases:
   1) C_2 5) C_4
   2) $D_{1,r}$ 6) D_4
   3) $D_{2,r}$ 7) C_6
   4) $D_{2,rh}$ 8) $D_{3,s}$

7. 1) Are C_4 and $D_{2,r}$ isomorphic as abstract groups?
   2) Are C_4 and $D_{2,r}$ unimodularly equivalent?

8. Explain why the following pairs of point groups are <u>not</u> unimodularly equivalent:
   1) $D_{2,r}$ and $D_{2,rh}$

2) C_6 and $D_{3,s}$

3) $D_{3,s}$ and $D_{3,\ell}$

9. Let G be a space group with translation subgroup T. Explain why, for any t ε T, the coset T(A,a) = T(A,a+t).

10. Verify that, for any s ε R^2, it is the case that
$$(I,s) (A,a)(I,s)^{-1} = (A, a - (A-I)s).$$

11. (cf. (7)). Let the space group G have (A_1, a_1), (A_2, a_2), . . . , (A_n, a_n) as coset representatives of T in G. Suppose there exists an s ε R^2 with
$$(A_i - I)s \equiv a_i \pmod 1, \text{ for all } i.$$
Prove that G is then equivalent to the trivial solution.

12. Suppose that the only vector x for which the square matrix A has Ax = x is x = 0. Prove that then the matrix A - I is invertible.

13. Explain why we may always assume that the element (I,0) is one of the coset representatives of T in G.

14. Show that the point group C_4 yields only the trivial solution.

15. Show that $D_{2,rh}$ yields only the trivial solution.

16. In (9), show that G_3 is not equivalent to G_1.

17. Find all space groups with point group D_4.

8 The RSA Public Key Cryptosystem

Since ancient times the making and breaking of secret codes
has been a vital concern in the military and diplomatic arena. More
recently, with the growing reliance on electronically stored or
transmitted data, the need to protect it from unwanted intrusion or
manipulation has prompted greater interest in the design of codes which
would provide a high degree of security. Because of this mathematicians
and computer scientists have expressed renewed interest in cryptography,
the study of coding.

It is not difficult to design an "unbreakable" code. If you
and I agree ahead of time that the word "fish" will mean "attack," an
enemy would have no a priori way of determining the meaning of the
message "fish." This is an example of a "one-time pad," a technique
used often by espionage agents. The problem with such a pad is inherent
in its name--it can, in essence, be used only once. The enemy, by
observing the consequences of its repeated use, could reasonably be
expected to deduce its meaning. Also, to send lengthier messages
elaborate dictionaries acting as keys would have to be constructed and
exchanged by the parties using the code. This would make the code bulky
and inefficient, and present the additional problem of distributing and
making secure these dictionaries. To be useful in practice, a code
should be relatively easy to implement--allow, say, rapid encoding and
decoding by computer. As importantly, it should permit repeated use
with minimal danger of being compromised and be able to withstand
prolonged attempts by computer to break it.

In this chapter we describe a mathematically based code
developed in 1977 by Ronald Rivest, Adi Shamir and Len Adleman which
attempts to meet the above criteria. Making use of the authors'
initials, we refer to it as the RSA system. The RSA code relies on
certain elementary properties of modular arithmetic and bases its
security on the difficulty of prime factorizing large numbers. Our
development is self-contained and begins with an exposition of the
algebraic ideas underlying the workings of the code.

Algebraic Preliminaries

The set of positive integers less than or equal to n that are relatively prime to n forms a group under multiplication (mod n). To be more specific,

<u>Def.</u>   For $n \geq 1$,

$$Z_n^* = \{x \mid 1 \leq x \leq n \text{ and } (x,n) = 1\}.$$

(Here, (x,n) denotes the greatest common divisor or gcd of x and n).

<u>Example:</u> $Z_4^* = \{1,3\}$

$Z_7^* = \{1,2,3,4,5,6\}$

Making use of the language of congruences we define a group operation on Z_n^* as follows:

if $x, y \in Z_n^*$, then

x • y is defined to be the <u>least residue</u> (mod n) of the ordinary integer product xy.

That is,

x • y = z, where $z \equiv xy$ (mod n)
and $1 \leq z \leq n$.

<u>Example</u>

In Z_4^*,   3 • 3 = 1

In Z_7^*,   3 • 3 = 2 and 6 • 4 = 3 .

Using elementary properties of congruences it is relatively straightforward to verify that the properties of closure and associativity hold for the above operation with the integer 1 serving as the identity.

The fact that each element of Z_n^* has a multiplicative inverse is a consequence of a very basic number theoretic fact which we record as a

<u>Lemma</u>   If x and n are integers that are relatively prime, then there always exist integers a and b with

$$ax + bn = 1.$$

Remarks

1. 3 and 7 are relatively prime and we can write
$$5 \cdot 3 + (-2) \cdot 7 = 1.$$

2. The above Lemma is proved in most all algebra and number theory texts and should have a familiar ring to those who have used the Euclidean Algorithm. Typical references are [9] and [6].

So that, if x is an element of Z_n^*, the Lemma guarantees the existence of integers a and b with

$$ax + bn = 1.$$

Considering the above statement (mod n), it translates to

$$ax \equiv 1 \ (\text{mod } n), \text{ since } bn \equiv 0 \ (\text{mod } n).$$

Letting a' be the least residue of a (mod n), we then have

$$a'x \equiv 1 \ (\text{mod } n), \text{ since } a' \equiv a \ (\text{mod } n).$$

But a' is an element of Z_n^* (cf. Exercise 3) and the above congruence then says that a' is the inverse of x in Z_n^* . For example, in Z_7^* , the inverse of 3 is 5.

So we get

Proposition Z_n^* is a group.

By its very definition, Z_n^* has order equal to the number of positive integers less than or equal to n that are relatively prime to n--a number often designated $\phi(n)$.

Def.   For $n \geq 1$,

$\phi(n)$ = the number of positive integers x with $1 \leq x \leq n$ and $(x,n) = 1$.

(ϕ is often referred to as the Euler-Phi-Function).

Examples

1. $\phi(8) = 4$
    $\phi(3) = 2$
    $\phi(7) = 6$

2. $|Z_{15}^*| = \phi(15) = 8.$

If p is a prime, then $\phi(p) = p - 1$, since all positive integers less than p are relatively prime to p. It is also true that

Lemma   If p and q are <u>distinct</u> primes, then $\phi(pq) = (p-1)(q-1)$.

Proof: One way of seing this is to observe that the integers relatively prime to the product pq are precisely those that are relatively prime to both p and q.

If we let

$$A = \{x \mid 1 \leq x \leq pq \text{ and } x \text{ is a multiple of } p\},$$

$$B = \{x \mid 1 \leq x \leq pq \text{ and } x \text{ is a multiple of } q\},$$

and $$X = \{x \mid 1 \leq x \leq pq\},$$

then the set of integers counted in determining $\phi(pq)$ is exactly the set $A' \cap B'$ (the prime denotes complementation relative to the set X).

By DeMorgan's Law,
$$A' \cap B' = (A \cup B)' \qquad \text{and thus}$$

$$|A' \cap B'| = |(A \cup B)'| = |X| - |A \cup B|$$

(1) $$= |X| - (|A| + |B| - |A \cap B|).$$

The very definitions of the sets involved show that
$$|A| = q, \ |B| = p, \ |A \cap B| = 1, \text{ and } |X| = pq.$$

Thus,

$$|A' \cap B'| = pq - (q + p - 1)$$
$$= (p - 1)(q - 1).$$

The proof ends with remarking again that $|A' \cap B'|$ is also, by construction, precisely $\phi(pq)$.

Actually, in general, given the prime factorization of an integer $n = P_1^{a_1} P_2^{a_2} \ldots P_k^{a_k}$, it can be shown that

$$\phi(n) = P_1^{a_1 - 1} (P_1 - 1) P_2^{a_2 - 1} (P_2 - 1) \ldots P_k^{a_k - 1} (P_k - 1) \text{ (cf. [6])}$$

Recall that a consequence of Lagrange's Theorem for Groups is that in any finite group G it is true that $X^{|G|} = 1$ for any element, X, of G (1 is the identity of G). In our situation, since $|Z_n^*| = \phi(n)$, this implies that

$$x^{\phi(n)} = 1 \text{ for any } x \text{ in } Z_n^*.$$

Stated in the language of congruences, this gives

121

<u>Proposition</u>   For any x with $(x,n) = 1$, $x^{\phi(n)} \equiv 1 \pmod{n}$.

<u>Remarks</u>

1. Since $\phi(21) = \phi(3 \cdot 7) = (3-1)(7-1) = 12$, we are guaranteed that, for example,
$$10^{12} \equiv 1 \pmod{21}.$$

2. The above Proposition is a classic result often referred to as Euler's Theorem or the Euler-Fermat Theorem.

A Description of the RSA System

Assume the existence of a person A to whom coded messages will be sent. We will describe how a sender encrypts a message to A and how, in turn, A decodes this message. A discussion of the security of this procedure follows.

Person A chooses three integers

	e	(used in encrypting messages to A)
	d	(used by A in decoding)
and	n	(the modulus),

with   (2)

	$n = pq$	where p and q are distinct primes,
	e	relatively prime to $\phi(n) = (p-1)(q-1)$
and	d	the multiplicative inverse of e in $Z_{\phi(n)}^*$ (i.e. with $de \equiv 1 \pmod{\phi(n)}$).

The reason for the technical constraints on these integers will be made clear shortly.

Messages M to be transmitted to A are first converted into digit form by some standard non-secret procedure. For example, we could use the assignment

$$A = 01$$
$$B = 02$$
$$\cdot$$
$$\cdot$$
$$\cdot$$
$$Z = 26$$
$$blank = 00,$$

and substitute a two-digit number for each letter. "HELLO" would then be written as "0805121215." So it does no harm to think of M as an integer.

To encrypt (send a message to A) we compute M^e (mod n). The encrypted message sent to A is the least residue of M^e (mod n).

Simple Example

$$n = 33 = 3 \cdot 11$$
$$e = 7$$

If the message M is just M = 19 then $M^e \equiv 19^7 \equiv 13$ (mod 33). So, "13" is transmitted.

Having received M^e (mod n), A then seeks to recover the original message M. The point is that the integer d allows A to do just this since it is the case that

$$(M^e)^d \equiv M \text{ (mod n)}.$$

This is the content of the

Lemma   If the integers e, d and n are chosen according to (2), then

$$M^{ed} \equiv M \text{ (mod n)}.$$

Proof: Since e and d satisfy

$$ed \equiv 1 \text{ (mod } \phi(n))$$

we can write $ed = k \phi(n) + 1$, for some integer k.

Now, n = pq. We first show that $M^{ed} \equiv M$ (mod p).

If $(M,p) = 1$, then since $\phi(p) = p - 1$, a previous Proposition (Euler's Theorem) says that $M^{p-1} \equiv 1$ (mod p). Thus,

$$M^{ed} \equiv M^{k \, \phi(n)+1} \equiv M^{k(p-1)(q-1)} M \text{ (mod p)}$$

$$\equiv (M^{p-1})^{k(q-1)} M \text{ (mod p)}$$

$$\equiv (1)^{k(q-1)} M \text{ (mod p)}$$

$$\equiv M \text{ (mod p)}.$$

In the event M is not relatively prime to p, then p must divide M, since p is a prime. So, $M \equiv 0$ (mod p) and any power of M must also be congruent to zero (mod p). Hence, any power of M is congruent to M itself. In particular, in this case also, $M^{ed} \equiv M \pmod{p}$.

The same argument, using the prime q instead of p, yields
$$M^{ed} \equiv M \pmod{q}.$$

Thus, the distinct primes p and q divide the quantity $(M^{ed} - M)$ -- _i.e._ it is true that $M^{ed} \equiv M \pmod{n}$. This conclusion could also have been reached using the Chinese Remainder Theorem discussed in Chapter 1.

The entire process can then be summarized in the diagram

$$M \xrightarrow{\text{encrypted}} M^e \pmod{n} \xrightarrow[\text{by A}]{\text{decrypted}} (M^e)^d \pmod{n} \xrightarrow{\text{yields}} M$$

original message   transmitted message   original message is recovered by A

Example

Continuing with the simple example used above where $n = 33 = 3 \cdot 11$ and $e = 7$, we note that d must be the multiplicative inverse of 7 in $Z_{\phi(n)}^* = Z_{20}^*$.

So, $d = 3$, since $3 \cdot 7 \equiv 1 \pmod{20}$. To decrypt the received message "13," A computes $13^d \equiv 13^3 \pmod{33}$. Note that $13^3 \equiv 19 \pmod{33}$, so the original message M = "19" has been recovered.

Remarks

1. It is important that the original message M be in the range $0 \le M \le n - 1$ (_i.e._ be a least residue mod n), since otherwise it would be impossible to distinguish it from any larger integer congruent to it (mod n). Longer messages that exceed this size are then broken into smaller blocks which are in the appropriate range and the message is then sent block by block. In practice, the modulus n is chosen to be large, on the order of 200 digits, so that block sizes up to 10^{200} can be used.

2. Several issues remain to be considered. What is the level of security provided by a code of this type? Can it be implemented efficiently? What advantages does it provide over traditional

cryptographic methods? To address these questions and get to the heart of what motivated the RSA system we turn briefly to the realm of prime numbers.

The Problem of Primes

The prime numbers can certainly be viewed as the building blocks of our familiar system of integers, but beyond their structural utility they seem to possess an inherent beauty and mystique that have fascinated mathematicians and amateurs for centuries. Questions about prime numbers are simultaneously often easy to phrase and grasp yet notoriously difficult to answer.

Two very basic and seemingly related questions are: Given a number n, 1) is n a prime number? and 2) what are the prime factors of n? In theory, 1) always admits a yes or no answer, yet that in practice this question is not trivial might be gleaned by pondering whether the 41 digit number $111 \cdots 11$ consisting of all 1's is a prime. Testing a number for primality by brute force (<u>i.e.</u> trial division) can be a costly and time-consuming process. Given a computer that could perform one million trial divisions per second, checking our test number would still require approximately 10^{14} seconds which is well over 1 million years. Since trial division is so lengthy, mathematicians have always been concerned with whether alternate tests for primality could be developed that would not necessarily force the user to eliminate smaller numbers one by one from being possible divisors. For example, if one could show that prime numbers have certain properties and if a number at hand lacked one of these properties, then certainly it could not be prime. One simple such test is based on Euler's Theorem. Say n were conjectured to be prime. Then, for any $0 < b < n$, it would be true by Euler that

$$b^{n-1} \equiv 1 \ (\text{mod } n).$$

So that, if we could find a b for which the above congruence is <u>not</u> valid, it would follow that n is <u>not</u> prime. <u>E.g.</u>, since the relation $5^8 \equiv 1 \ (\text{mod } 9)$ is <u>not</u> true, we may conclude that 9 is not prime.

With the greater availability of high speed computers more elaborate versions of such tests have recently been developed that allow the computer testing for primality of heretofore unmanageable numbers in

reasonable amounts of time. Good accounts of this progress are
available in [13] and [5]. As of this writing a 100 digit number can be
tested for primality in about 40 seconds, while certifying that a 200
digit number is or is not prime can be accomplished in approximately 10
minutes. One byproduct of this recent work is that it is now relatively
easy to produce fairly large prime numbers. If a 100 digit prime number
were desired, one could generate random 100 digit numbers, test them for
primality using the newer available methods, stopping when a prime is
found. Probabilistic arguments which we do not include can be given to
show that such a search would terminate fairly quickly.

The status of the second question -- What are the prime
factors of n? -- is somewhat different. While many algorithms for
factorization of an integer n have been developed, none of them is even
close in speed to the primality testing routines discussed above.
Currently the fastest known algorithm for factoring an integer n
requires approximately $e^{\sqrt{\ln(n)\,\ln(\ln n)}}$ steps (cf. [3], p. 31).
A feeling for the efficiency of this algorithm can be obtained by
assuming a computer that carries out 1 million steps per second and
estimating the amount of time the algorithm would require in terms of
the size of the number n being tested.

No. of Digits in n	No. of Steps in Algorithm	Running Time of Algorithm
50	1.4×10^{10}	3.9 hours
100	2.3×10^{15}	74 years
200	1.2×10^{23}	3.8×10^9 years

Thus, while we can reasonably expect to declare in a short
time whether a 100 digit number is prime, it does not seem reasonable to
expect to be able to factor it using current techniques. The outlook
for factorization of a 200 digit number is much worse.

The point is that the questions originally posed at the beginning of this section are not equivalent. It is not necessary to try to factor a number to see whether it is prime or composite. Given that it has been found to be composite, asking also for its factors is asking a seemingly much more difficult question. There are some mathematicians who believe that the problem of factoring will be shown to be intrinsically difficult. That is, they believe that efficient factorization algorithms will never be found. It is this suspicion that factoring is very hard that underlies the construction of the RSA cryptosystem and led Martin Gardner in the Scientific American to call it "a new kind of cipher that would take millions of years to break" [7].

Features of the RSA System

We return to our person A who has selected integers e, d, and n = pq according to the requirements of (2).

The integers e and n are made <u>public</u>, while the decryption integer d and the primes p and q which are the factors of n are kept <u>secret</u>.

The reason for making e and n public is to allow <u>anyone</u> to send a message M to A: such a sender would simply transmit M^e (mod n).

The integer d clearly must be kept secret. Otherwise, any enemy with knowledge of d could decipher a transmitted message since $(M^e)^d \equiv M$ (mod n).

The reason the primes p and q which determine n are also kept secret is that if an eavesdropper knew p and q, then he could easily compute $\phi(n) = (p-1)(q-1)$. But then, since e and n are already public, he could discover d since d is determined by $de \equiv 1$ (mod $\phi(n)$).

Rivest, Shamir and Adleman suggest that the integer n be constructed by first finding two large random primes of approximately 100 digits each and then forming n by taking the product pq of these primes. Again, only the 200 digit product n is announced publicly, not p or q.

How could such an RSA cryptosystem be broken? Certainly, if

the factors p and q were found then, as shown above, d would be known and the code broken. But the point is that the problem of factoring a 200 digit number is well beyond the reach of existing algorithms and computing methods. The state of the art in factorization methods finds factoring such a number out of the question and the authors of the RSA system (and others) do not think that future developments will pose much of a threat. Factoring large numbers seems too hard a problem.

Another way to crack the code would be to determine $\phi(n)$ — since then d could be found by solving the congruence $xe \equiv 1 \pmod{\phi(n)}$ for x. But RSA argue that this is no easier than attempting to factor n; if one knew $\phi(n)$ then p and q could be found using the following argument.

$$
\begin{aligned}
\text{Since } \phi(n) \ &= (p - 1)(q - 1) \\
&= pq - (p + q) + 1 \\
&= n - (p + q) + 1,
\end{aligned}
$$

knowledge of $\phi(n)$ would then yield knowledge of $p + q$.

Also, the relation

$$
\begin{aligned}
(p - q)^2 \ &= (p + q)^2 - 4pq \\
&= (p + q)^2 - 4n,
\end{aligned}
$$

shows that, knowing $p + q$ and n, one could find $p - q$. But then p and q are found from

$$
p = \frac{1}{2} [(p + q) + (p - q)],
$$

$$
q = \frac{1}{2} [(p + q) - (p - q)] .
$$

Thus, any attempt to find $\phi(n)$ would be equivalent to solving the hard problem of factoring n.

Could the decrypting integer d be found some other way, or could the RSA system be broken without finding d? We return to this question in our concluding remarks.

Implementation of an RSA system requires the ability to be able to compute expressions of the form $M^e \pmod{n}$ efficiently and rapidly. Note that the decoding procedure involves an analogous computation using d instead of e.

128

That this can be done efficiently can be seen by the observation that to compute, say M^{13}, does not require 12 or 13 multiplications; rather, since

$$M^{13} = M^{(8+4+1)} = (M^4)^2 \cdot (M^2)^2 \cdot M,$$

M^{13} can be computed using only 5 multiplications. This is clearly connected with the fact that $13 = 2^3 + 2^2 + 2^0$; _i. e._ that 13 has the binary representation 1101. An algorithm computing M^e (mod n) could then take advantage of the binary representation of e, computing and multiplying together those powers of M that correspond to the powers of 2 that occur in the representation. At each stage the computations could be reduced (mod n).

An example of such an algorithm is given below. To give it a cleaner appearance, the reduction of the computations (mod n) is omitted. The final value of Z is the desired M^e.

(3)   Algorithm to compute M^e

1.   Find the binary representation $e_K \, e_{K-1} \ldots e_1 \, e_0$ of e

2.   Let X = M
     If e_0 = 1 Let Z = M else let Z = 1

3.   For I = 1 to K
         X = X * X
         If e_i = 1 Let Z = Z * X

     NEXT I

4.   Return Z

        The number of multiplications required by this algorithm is never greater than 2 $\log_2 e$ (cf. Exercise 13). So that even when e is a 200 digit number, the computation of M^e will require at most $2 \log_2 10^{200} = 400 \log_2 10 \approx 1,330$ multiplications, instead of a naive 10^{200} brute force multiplications.

A Better Example

We take n = 47 · 59 = 2773 and e = 17. Then $\phi(n)$ = 46 · 58 = 2668 and the decryption key d is found by solving the congruence
$$17x \equiv 1 \pmod{2668}.$$
This yields d = 157.

Using the correspondence introduced in the earlier example, the phrase

"ABSTRACT ALGEBRA"

is represented by

0102   1920   1801   0320   0001   1207   0502   1801.

We encode this using blocks of length 4. The first block M = 0102 is encrypted by computing
$$M^{17} \equiv 102^{17} \equiv 1395 \pmod{2773}.$$
Thus "1395" is the transmission representing the first four digits of the message.

The entire encrypted message is

1395   2109   2003   2299   0001   0723   1804   2003.

As a check on the decryption process we compute
$$(1395)^{157} \pmod{2773}$$
and find that its least residue is indeed the original block 0102.

It is not difficult to write computer programs that do the above computations, provided care is taken to ensure that the integer arithmetic is carried out exactly.

Public Keys and Signatures

Think of our person A to whom messages were being sent as being the proprietor of the integers e, d, and n specified in (2) -- so we can index these integers as belonging to A: e_A, d_A, and n_A. In a system where secret messages are being sent among many users, _e.g._, a telex network, each user in the network would then likewise be an owner of his own unique triple of integers -- user B would thus possess e_B, d_B, n_B. Each user would make public his e and n, allowing anyone else in the network to encrypt and send a message to him. But, to prevent any unauthorized listening, each person would keep his decoding integer d secret -- only the intended person can decode and properly receive messages sent to him. Such a network would then have the novel feature that anyone would have the means to send a message to anyone else. The details of the encoding procedure are open knowledge -- each person's e and n could, for instance, be listed in a public directory, much like a telephone book. This can be done because, as noted previously, the prospect that d can be recovered from e and n is exceptionally remote. Since half its workings are revealed openly, the RSA system is often referred to as a public-key-cryptosystem.

A primary area of application for such systems is the electronic transmission of data, where the dual concerns of speed and secrecy often occur. In such cases not only is it important that the content of messages be kept secret, but also that there be some way of verifying that the person sending the message is indeed who they claim to be. For example, were a coded authorization sent to the Bank of England to transfer ten million pounds, the bank would be anxious to assure itself of the identity of the sender. This is often referred to as the problem of authentication. We describe how the RSA system provides senders a way of signing their messages that would convince receivers of their legitimacy.

Given a message we let $E_A(M)$ be the result of encrypting the message M using A's public e_A and n_A. That is, $E_A(M)$ is the least residue of $M^{e_A} \pmod{n_A}$. Given a received transmission R, $D_A(R)$ denotes the result of A's decrypting R using d_A -- $D_A(R)$ is the least residue of $R^{d_A} \pmod{n_A}$. Note that

(4)
$$D_A(E_A(M)) = M$$

and
$$E_A(D_A(M)) = M \quad .$$

Now, suppose A sends a transmission to B and wants to convince B that it is indeed A that is sending the message. A constructs a digital message S which serves as a signature. He then first computes $D_A(S)$ and then $E_B(D_A(S))$. It is the message $E_B(D_A(S))$ that is transmitted. B receives this message and applies his secret decoding key to it: this yields $D_B(E_B(D_A(S))$, which by (4), equals $D_A(S)$.

B now applies E_A (which is publicly available) to $D_A(S)$ and by (4), produces $E_A(D_A(S)) = S$. B has now received A's signature and is assured that it must have been A sending it -- no one else but A would know D_A. Put another way, if a person other than A had sent the message, that person would have been forced to use, say, D_C for lack of knowing D_A. The sent message would be $E_B (D_C (S))$. B's decryption, as before, produces $D_C (S)$, but then when D attempts to apply E_A to this, it will <u>not</u> be the case that $E_A (D_C (S)) = S$. The reason is that e_A is not the multiplicative inverse of d_C (mod n_A) -- by design, e_A is the multiplicative inverse of d_A and the decryption keys d_A and d_C are distinct. Thus the RSA system does not permit someone else to masquerade as A.

Concluding Remarks

The strength of the RSA system rests on the perception that factoring large integers is and will remain a very hard problem. It has been shown that uncovering the secret decryption key d is equivalent to factoring the modulus n. But it could possibly be the case that the RSA scheme could be broken <u>without</u> finding d and <u>without</u> factoring n -- though this has not been done. There is no proof available that <u>any</u> attempt at breaking RSA is necessarily equivalent to finding an efficient factoring algorithm. Further, there is no proof available that factoring is by its very nature an intractable problem. Just recently (February, 1984) mathematicians at Sandia National Laboratories in Albuquerque announced that they had successfully factored a 69 digit number in the astoundingly short time of 32 hours (cf. [17]). Though this is a far cry from factoring a 200 digit number, it does provide a healthy reminder of the fact that a problem thought hopelessly hard by many can at times yield to a surprise clever attack. On the other hand, in their 1977 report announcing the system, the authors of RSA challenged others to find a way to break their method -- as of this writing the system has withstood attempted attacks on it.

If encryption and decryption are viewed as inverse operations or functions, the RSA system has the feature that, while both operations are computationally feasible, knowledge of the encryption algorithm does not provide an easy way of discovering the inverse decryption algorithm. Diffie and Hellman [4] have characterized systems that depend on operations where knowledge of the function in one direction does not betray an easy method for computing its inverse as "trap-door

132

systems" -- easy to fall into, but difficult to get out of. In 1977 Hellman and his student, Ralph Merkle, proposed their own example of such a system, based on a famous problem in operations research called the knapsack problem. Their system, known as the trapdoor knapsack, seemed also to provide an example of an "unbreakable" public-key-cryptosystem. Yet in 1982, Adi Shamir, one of the authors of the RSA system, broke the original version of the code developed by Merkle-Hellman (cf. [10]). So far, Hellman and Merkle have been unable to return the favor, but it is perhaps too early to tell whether the RSA system will stand the test of time.

The survey article [4] by Diffie and Hellman and the monograph [3] provide good backgrounds to contemporary cryptography. Descriptions of the RSA system are contained in [8], [3], [11], [1], and [16]. The original paper [15] by Rivest, Shamir and Adleman is easy to follow and worth reading. Good commentaries on the problems of primality testing and factoring are contained in [12], [5], [14], and [13].

References

[1] H. Beker and F. Piper

Cipher Systems, John Wiley and Sons, 1982

[2] G.R. Blakely and I. Borosh

"Rivest - Shamir - Adleman Public Key Cryptosystems Do Not Always Conceal Messages", Comp. & Maths. with Appls., Vol. 5, pp. 169-178.

[3] R.A. DeMillo, George I. Davida et al.

Applied Cryptology, Cryptographic Protocols, and Computer Security Models, Proceedings of Symposia in Applied Mathematics, Vol. 29, American Mathematical Society, 1983

[4] Whitfield Diffie and Martin Hellman

"Privacy and Authentication: An Introduction to Cryptography", Proceedings of the IEEE, Vol. 67, No. 3, March 1979, pp. 397-427

[5] John D. Dixon

"Factorization and Primality Tests," American Mathematical Monthly, Vol.

91, No. 6, June-July 1984
pp. 333-352.

[6] Underwood Dudley

Elementary Number Theory
(Second Edition), W.H.
Freeman, 1978

[7] Martin Gardner

"A New Kind of Cipher that
Would Take Millions of Years
to Break," Mathematical
Games Section of Scientific
American, August 1977, pp.
120-124.

[8] Martin E. Hellman

"The Mathematics of Public-
Key Cryptography,"
Scientific American, August
1979, pp. 146-157.

[9] A.P. Hillman and G.L. Alexanderson

A First Undergraduate Course
in Abstract Algebra (Third
Edition), Wadsworth, 1983

[10] Gina Kolata

"New Code is Broken"
Science, Vol. 216, May 28,
1982, pp. 971-2

[11] Alan G. Konheim

Cryptography: A Primer,
John Wiley and Sons, 1981

[12] Carl Pomerance

"Recent Developments in
Primality Testing,"
Mathematical Intelligencer,
Vol. 3, No. 3, January 1982,
pp. 97-105.

[13] Carl Pomerance

"The Search for Prime
Numbers", Scientific
American, December 1982, pp.
136-147.

[14] Carl Pomerance

Lecture Notes on Primality
Testing and Factoring, MAA
Notes Number 4, Mathematical
Association of America,
1984.

[15] R.L. Rivest, A. Shamir, and L. Adleman "A Method for Obtaining
Digital Signatures and
Public-Key Cryptosystems",
Communications of the ACM,
February 1978, Vol. 21, No.
2, pp. 120-126.

[16] Gustavus J. Simmons   Cryptology: The Mathematics
                                        of Secure Communication,
                                        Mathematical Intelligencer,
                                        Vol. 1, No. 4, 1979, pp.
                                        233-246.

[17] _____   "Cracking a Record Number"
                                        Time Magazine, February 13,
                                        1984, p. 47.

Exercises

1. List the inverses of all the elements of

    1) Z_{10}^*

    2) Z_{12}^*

2. Is Z_5^* cyclic? Is Z_8^* ? Is Z_{10}^* ?

3. Suppose $x \in Z_n^*$ and there exist integers a and b with $ax + bn = 1$.
   Let a' be the least residue of a (mod n). Explain why $a' \in Z_n^*$.

4. Verify the set counting argument used in (1).

5. Suppose p is a prime and $(a,p) = 1$.
   Prove: $a^{p-1} \equiv 1$ (mod p).

6. In the RSA system, suppose n is chosen as $n = 65$. Find the
   decryption key d for the following choices of e.
                   1) $e = 5$
                   2) $e = 7$
                   3) $e = 11$

7. Let $n = 33$ and $e = 3$.
   1) Encrypt the following digital messages (divide into blocks of
      two).
                   a) M = 18
                   b) M = 1801
                   c) M = 021319
                   d) M = 11002416

   2) Find the decryption key d corresponding to the above choice of e
      and <u>decrypt</u> the following messages (divide into blocks of two).
                   a) M = 06
                   b) M = 0612
                   c) M = 231416
                   d) M = 17000614

8. Use the correspondence A = 01, B = 02, ..., Z = 26, blank = 00 to
   convert the following messages to digital form.
                   1) "BYE"
                   2) "GROUP"

3) "LOVE YOU"

9. With modulus n = 35 and e = 5, encrypt the digital messages in the
    above exercise (divide into blocks of two).

10. Suppose A has made public n = 91 and e = 5. Assume you do not know
    how to factor the modulus n, but that you are able to steal the
    information that $\phi(n) = 72$. Explain how you could now break A's
    code.

11. What would happen if the RSA system were designed with n a prime,
    instead of the product of two primes? Would it then be easily
    breakable?

12. How many multiplications does the algorithm (3) use to compute
    M^{45}? to compute M^{64}?

13. Explain why the number of multiplications required by algorithm (3)
    to compute M^e never exceeds $2\log_2 e$.

14. Explain why $E_A(D_A(M)) = M$ if M is a digital message in the range
    $0 \le M \le n_A - 1$.

15. Look up the paper [15] and read it.

16. This exercise shows that in certain instances RSA encryption may
    leave messages unaltered.
    1) Let n = 15, e = 3.
        How many X are there, $0 \le X \le n - 1$, with $X^e \equiv X \pmod{n}$?
    2) Same question, with n = 15 and e = 5.

17. In this exercise, given n = pq, p and q distinct primes, and e, with
    $(e, \phi(n)) = 1$, we count the number of messages X with $X^e \equiv X \pmod{n}$.

    1) Show that $Z_n^* \simeq Z_p^* \times Z_q^*$

    2) Show that $X^e \equiv X \pmod{n}$ if and only if $X^e \equiv X \pmod{p}$ and
        $X^e \equiv X \pmod{q}$.

    3) Prove that $X \in Z_p^*$ has $X^e = X$ if and only if the order of X
        divides e - 1.

    4) How many elements of Z_p^* have order dividing e - 1?
        (Hint: Let t = gcd of e - 1 and p - 1
                      = (e - 1, p - 1).

Elements of order dividing e - 1 must then be in the unique subgroup of order t in Z_p^* . For the purpose of this exercise, you may assume the fact that if p is prime, Z_p^* is cyclic (cf. [6], p. 79)).

5) Prove that if $(x, p) \neq 1$, then $X^e \equiv X$ (mod p) if and only if $X \equiv 0$ (mod p).

6) Conclude that the number of X with $0 \leq X < p$ and $X^e \equiv X$ (mod p) is $1 + (e - 1, p - 1)$.

7) Use 1), 2) and 6) to show that the number of X with $0 \leq X < n$ and $X^e \equiv X$ (mod n) is

$$[1 + (e - 1, p - 1)] [1 + (e - 1, q - 1)].$$

8) Compare the result in 7) to your answers to Exercise 16.

137

9 Integer Programming

Integer programming deals with the problem of maximizing or minimizing certain expressions with the restriction that the quantities involved must be integers. The problem, for example, of what is the largest that $2x_1 - x_2$ can be if x_1 and x_2 are integers with $0 \leqslant x_1 \leqslant x_2$ and $0 \leqslant x_2 \leqslant 4$ is a simple example of what is called an integer program. Much of the interest paid to integer programming is due to the fact that many "real world" problems can be expressed as integer programs. Examples include the scheduling of airline crews, capital budgeting, optimal loading of a freighter and the famous Traveling Salesman Problem [2]. A glance at a bibliography of the sort [11] will reveal the wealth of applications of integer programming.

Integer programming is closely related to linear programming which addresses the same question of optimization, without the restriction to integer variables.

$$
\begin{aligned}
&\text{Maximize} &&2x_1 - x_2 \\
&\text{subject to} &&0 \leqslant x_1 \leqslant x_2 \\
& &&0 \leqslant x_2 \leqslant 4
\end{aligned}
$$

(x_1, x_2 need not be integers)

is an example of a linear program. In practice most linear programs can be solved by use of the simplex method, a technique advanced by George Dantzig in 1947.

Hence, to solve an integer program (IP) the temptation exists to treat it as a linear program (LP) (i.e., ignore the integer constraints), solve the LP, and then in some sense "round the solution

138

off" to the nearest integer to produce an optimal solution to the IP. Sadly, this does not work in general.

Example

The LP problem

$$\text{Maximize} \quad M = 3x_1 + 4x_2$$

subject to

(1)
$$2x_1 + 3x_2 \leq 21$$
$$4x_1 + x_2 \leq 22$$
$$x_1 \geq 0, \quad x_2 \geq 0$$

has the optimal solutions $x_1 = 4\ 1/2$, $x_2 = 4$. The "rounded" values $x_1 = 4$ and $x_2 = 4$ give $M = 28$. But this is not the optimal solution to (1) treated as an IP, since $x_1 = 3$, $x_2 = 5$ satisfy the inequalities above and give a value of $M = 29$. Note that the values $x_1 = 5$, $x_2 = 4$ do not even satisfy the inequalities.

The connection between the solution to an IP and its solution considered as a LP is more subtle. In a series of investigations begun in 1958 Ralph Gomory of IBM explored this connection. In the process he showed that any given integer program has a finite abelian group attached to it and that solving the IP can often be reduced to solving relations in that finite group.

The purpose of this section is to outline this use of group theory in integer programming. Some minimal familiarity with elementary linear programming including use of simplex tableaux is assumed. The necessary background can be readily obtained from the first few chapters of texts such as [5] or [9].

139

Background

Discussion of linear and integer programming problems is greatly simplified by using the language of matrices and vectors which we now introduce.

We treat LP problems of the form

Maximize $\quad$ CX

with $\qquad$ $AX = b, \quad X > 0$

where $\qquad$ A is an m x n matrix of integers,

$$m < n$$

b is an m x 1 vector of integers

(2) $\qquad$ C is a 1 x n vector

$$X = \begin{pmatrix} x_1 \\ x_2 \\ \vdots \\ x_n \end{pmatrix}$$

$X > 0$ means that $x_i > 0$, for all i.

Example

With the introduction of the slack variables x_3 and x_4 (1) can be expressed in the form above with

$$A = \begin{pmatrix} 2 & 3 & 1 & 0 \\ 4 & 1 & 0 & 1 \end{pmatrix}$$

$$b = \begin{pmatrix} 21 \\ 22 \end{pmatrix}$$

$$C = (3 \quad 4 \quad 0 \quad 0)$$

$$X = (x_1, x_2, x_3, x_4)$$

Note that technically we should write X as a column vector so that CX is a well-defined scalar and the product AX makes sense. At times, though,

140

to simplify the appearance of certain expressions we will take the liberty of writing X as a row vector.

Remark

We also make the assumption that A has rank m and contains the $m \times m$ identity matrix as a submatrix. This is tantamount to assuming that the constraints in (2) arise from inequalities.

If we permute columns so that A is written in the form

$$A = (B : N)$$

where B is a non-singular $m \times m$ matrix and correspondingly permute the components of X and C to write them in the form

$$X = (X_B, X_N)$$
$$C = (C_B, C_N)$$

then problem (2) becomes

Maximize $\qquad C_B X_B + C_N X_N$

with $\qquad BX_B + NX_N = b,$
$$X_N, X_B \geq 0.$$

The variables X_B are called <u>basic variables</u> corresponding to the basis B, the variables X_N <u>non-basic</u> variables.

Since B is invertible, we have

$$X_B = B^{-1}b - B^{-1}NX_N,$$

and

$$C_B X_B + C_N X_N = C_B(B^{-1}b - B^{-1}NX_N) + C_N X_N$$

So, problem (2) becomes

Maximize $\qquad C_B B^{-1}b + (C_N - C_B B^{-1}N)X_N$

(3) $\qquad$ with $\qquad X_B = B^{-1}b - B^{-1}NX_N,$

141

$$X_N, \; X_B \geqslant 0.$$

<u>Example</u> In (1), if we choose

$$B = \begin{pmatrix} 2 & 1 \\ 4 & 0 \end{pmatrix}$$

then $$X_B = (x_1, \; x_3)$$

(4) $$X_N = (x_2, \; x_4)$$

$$C = (C_B, \; C_N) = (3 \quad 0 \quad 4 \quad 0)$$

$$N = \begin{pmatrix} 3 & 0 \\ 1 & 1 \end{pmatrix}.$$

The following observations are made in a development of elementary linear programming:

1) If $B^{-1}b > 0$ letting $X_N = 0$ and thus $X_B = B^{-1}b$ yields a basic feasible solution.

2) If $B^{-1}b > 0$ and also $C_N - C_B \, B^{-1}N < 0$ then the LP (2) has the <u>optimal</u> solution $X_N = 0$, $X_B = B^{-1}b$, since by (3) making any component of X_N greater than zero can only decrease (or leave the same) the value of the objective function.

3) The simplex algorithm can be used to locate such an optimal solution with corresponding basis B.

We adopt the following commonly used compact version of the

142

simplex tableau which gives an expression of the basic variables

$$x_B = \left(x_{B_1}, \; x_{B_2}, \; \ldots, \; x_{B_m}\right)$$

in terms of the <u>negatives</u> of the non-basic variables

$$x_N = (\ldots, \; x_j, \; \ldots, \; x_k, \; \ldots).$$

Non-basic variables

		$-x_j$	$\cdots$	$-x_k$	$\cdots$
M		$y_{00} \cdots y_{0j} \cdots y_{0k} \cdots$			
		$\vdots$ $\vdots$ $\vdots$			
Basic variables	x_{B_i}	$y_{i0} \cdots y_{ij} \cdots y_{ik} \cdots$			
		$\vdots$ $\vdots$ $\vdots$			
	x_{B_m}	$y_{m0} \cdots y_{mj} \cdots y_{mk} \cdots$			

In the above tableau,

$$y_{00} = C_B \, B^{-1} b$$

$$(y_{01}, \; \ldots, \; y_{0k}, \; \ldots) = C_B \, B^{-1} N - C_N$$

$$\begin{pmatrix} y_{10} \\ \vdots \\ y_{m0} \end{pmatrix} = B^{-1} b$$

$$\begin{pmatrix} y_{11} & \cdots & y_{1j} & \cdots \\ \vdots & & & \\ \vdots & \cdots & y_{ij} & \cdots \\ y_{m1} & \cdots & y_{mj} & \cdots \end{pmatrix} = B^{-1} N$$

(Note: This is the version used in [4]).

<u>Example</u>

The tableau corresponding to the choice of basic variables in

(4) is

143

		$-x_2$	$-x_4$
M	$\frac{33}{2}$	-3 1/4	3/4
x_1	5 1/2	1/4	1/4
x_3	10	2 1/2	-1/2

The associated group problem

The integer programming (IP) problem is problem (2) along with the additional constraint that the vector X have all integer coordinates. That is,

(5)

Maximize $\quad\quad\quad\quad\quad\quad\quad\quad$ M = CX

subject to $\quad\quad\quad\quad\quad\quad\quad$ AX = b

$$X > 0, X \text{ integer.}$$

Now for __any__ basis B, by (3) the objective function M can be written as $M = C_B \, B^{-1} b + (C_N - C_B \, B^{-1} \, N) \, X_N$. Since $C_B \, B^{-1} b$ is a __constant__, the above problem is then equivalent to

(IP)

Maximize $\quad\quad\quad\quad\quad$ $(C_N - C_B B^{-1} \, N) \, X_N$

with $\quad$ $X_B + B^{-1} \, N \, X_N = B^{-1} \, b$

$\quad X_B, \quad X_N > 0, \quad X_B, \, X_N$ integer.

Since Gomory's reduction of (IP) to a problem in groups involves considering "fractional parts" of vectors in the above constraint, we digress a moment to make this more precise.

Let R^m and Z^m denote the additive groups of m- dimensional real and integer vectors respectively. Then the quotient group R^m/Z^m can be thought of as the group of all m- dimensional real vectors $X = (x_1, \ldots, x_i, \ldots, x_m)$, $0 \leq x_i < 1$, where addition is performed (mod 1).

For example, with m = 2,

$$\begin{pmatrix} 1/3 \\ 3/4 \end{pmatrix} + Z^2 + \begin{pmatrix} 2/3 \\ 3/4 \end{pmatrix} + Z^2$$

$$= \begin{pmatrix} 0 \\ 1/2 \end{pmatrix} + Z^2.$$

The projection $f : R^m \longrightarrow R^m/Z^m$ by $f(X) = X + Z^m$ is a homomorphism of additive abelian groups. We will often write $\bar{X}$ for $f(X)$, where $X \in R^m$.

Let I be the index set for the variables occurring in X_N, i.e. corresponding to the columns of N. Let α_j, $j \in I$, denote the columns of $B^{-1}N$.

Then

$$B^{-1}NX_N = \sum_{j \in I} \alpha_j x_j .$$

Now, suppose $X = (X_B, X_N)$ is a feasible solution to IP. Then $X_B + B^{-1}NX_N = B^{-1}b$ and, applying the homomorphism f, we get

$$\overline{X_B} + \overline{B^{-1}NX_N} = \overline{B^{-1}b}.$$

Since X_B has __integer__ components, $\overline{X_B} = 0$. So, denoting by α_0 the column vector $B^{-1}b$, it follows that

$$\overline{B^{-1}N} X_N = \overline{B^{-1}b}$$

or

$$\sum_{j \in I} \bar{\alpha}_j x_J = \bar{\alpha}_0 .$$

We have thus shown that every feasible solution to IP is also a feasible solution to

Maximize $(C_N - C_B B^{-1}N) X_N$

(GP)   subject to $\sum\limits_{j \in I} \bar{\alpha}_j x_j = \bar{\alpha}_0$

$x_j > 0$, integral, $j \in I$.

We indicate why the above is referred to as a group problem (GP). The group in question is the additive abelian group generated by the $\bar{\alpha}_j$, $j \in I$ -- that is, the group generated by the fractional parts of the columns of $B^{-1}N$. We designate it by G_B. So that

G_B = the subgroup of R^m/Z^m generated by $\bar{\alpha}_j$, $j \in I$,

and the constraint in GP can be viewed as finding ways of expressing the group element $\bar{\alpha}_0$ in terms of the generators $\bar{\alpha}_j$.

A bound on the exponent of G_B can be obtained by noting that the elements of $B^{-1}N$ can be written as rational numbers with denominator D, where $D = |\det B| = $ absolute value of the determinant of B. This is so since $B^{-1} = \frac{1}{\det B}$ adj B and the matrices B and N have integer entries. Hence, $D \bar{\alpha}_j = 0$, for all $j \in I$, and so G_B has exponent $< D$. Since G_B is finitely generated abelian it follows that G_B is finite. We will shortly determine the actual order of G_B.

Thus $\sum\limits_{j \in I} \bar{\alpha}_j x_j = \bar{\alpha}_0$, $x_j > 0$, integer ,

can be viewed as a constraint in a finite abelian group. Our previous work shows that if IP is consistent, then GP is consistent.

Example

The group problem corresponding to the choice of basis B and

146

tableau in Example (4) is

$$\text{Maximize} \qquad\qquad 3\tfrac{1}{4}\, x_2 - \tfrac{3}{4}\, x_4$$

$$\text{subject to} \qquad \overline{\begin{pmatrix} 1/4 \\[4pt] 2\tfrac{1}{2} \end{pmatrix}} x_2 + \overline{\begin{pmatrix} 1/4 \\[4pt] -1/2 \end{pmatrix}} x_4 = \overline{\begin{pmatrix} 5\tfrac{1}{2} \\[4pt] 10 \end{pmatrix}}$$

$$x_2,\ x_4 \ \geqslant\ 0,\ \text{integers}.$$

Since we are dealing with fractional parts, note that the constraint can be written as

$$\overline{\begin{pmatrix} 1/4 \\[4pt] 1/2 \end{pmatrix}} x_2 + \overline{\begin{pmatrix} 1/4 \\[4pt] 1/2 \end{pmatrix}} x_4 = \overline{\begin{pmatrix} 1/2 \\[4pt] 0 \end{pmatrix}}.$$

Here, G_B is generated by $\overline{\begin{pmatrix} 1/4 \\ 1/2 \end{pmatrix}}$ and is cyclic of order 4.

The connection between GP and IP

We now address the questions of how the group problem GP may be useful in solving the original integer problem IP.

Suppose B is an optimal basis for IP considered as a linear program - $\underline{i.e.}$ ignore the integrality constraints. B can be found by the simplex method. If $X = (X_B,\ X_N)$ is integral, the integer program is then also solved. If not, consider the group problem corresponding to the basis B.

Since B is optimal, we are guaranteed by the theory of the simplex method that $(C_N - C_B\, B^{-1} N) \leqslant 0$. Thus the group problem

$$\text{Maximize} \qquad\qquad (C_N - C_B\, B^{-1} N)\, X_N$$

(GP)

subject to
$$\sum_{j \in I} \overline{\alpha}_j x_j = \overline{\alpha}_0$$

$$x_j \geqslant 0 \text{ , integer}$$

has an optimal solution. Intuitively, the x_j's should be chosen as small as possible, while still meeting the group constraint. So that an optimal solution to GP will certainly have

$$0 \leqslant x_j < \text{order of } \overline{\alpha}_j \leqslant D = \det B .$$

We are thus guaranteed that at worst the group problem can be solved via finite enumeration.

Suppose then that $X_N^* = (x_j^*)$ provides an optimal solution to GP. Define X_B^* by $X_B^* = B^{-1}b - B^{-1}N X_N^*$. X_B^* is then a vector with integer coordinates since $\overline{X}_B^* = \overline{B^{-1}b} - \overline{B^{-1}N} X_N^*$

$$= \overline{\alpha}_0 - \sum_{j \in I} \overline{\alpha}_j x_j^* = 0 .$$

Then, $X^* = (X_B^* , X_N^*)$ is an integer vector satisfying the constraint

$$B X_B + N X_N = b \text{ of IP.}$$

The only problem is that there is no a priori guarantee that the vector X_B^* satisfies $X_B^* \geqslant 0$ (cf. Example (7) following). But, in the event that it is indeed true that $X_B^* \geqslant 0$, we claim that the vector X^* constructed above provides an optimal solution to the original IP. The reason relies on the fact that the objective functions in IP and GP are identical and that any solution of IP provides a vector satisfying the constraints of GP. For, if $Y = (Y_B, Y_N)$ were a non-negative vector yielding a larger objective value in IP than X^*, then our results above show that Y_N satisfies the constraints of GP and yields a larger objective value than X_N^* -- contradicting the optimality of X_N^* . We summarize our discussion in

148

<u>Proposition</u>

Let B be an optimal basis for IP considered as a linear program. Let X_N^* provide an optimal solution to

Maximize $\qquad\qquad (C_N - C_B B^{-1}N)\, X_N$

with $\qquad\qquad\qquad \sum_{j\in I} \overline{\alpha}_j\, x_j = \overline{\alpha}_0,$

$\qquad\qquad\qquad\qquad X_N \geq 0,$ integer,

Let $X_B^* = B^{-1}b - B^{-1}NX_N^*.$

Then, if $X_B^* \geq 0,$

the vector $X^* = (X_B^*,\ X_N^*)$ gives an optimal solution to IP.

<u>Example</u>

An optimal tableau for the problem in Example (1) is

		$-x_3$	$-x_4$
M	29 1/2	$\frac{13}{10}$	$\frac{1}{10}$
x_1	4 1/2	$-\frac{1}{10}$	$\frac{3}{10}$
x_2	4	$\frac{2}{5}$	$-\frac{1}{5}$

(6)

The corresponding group problem becomes

Maximize $\qquad -\frac{13}{10}\, x_3 - \frac{1}{10}\, x_4$

subject to $\quad \left(\begin{array}{c} \overline{-\frac{1}{10}} \\ \frac{4}{10} \end{array}\right) x_3 + \left(\begin{array}{c} \overline{\frac{3}{10}} \\ -\frac{2}{10} \end{array}\right) X_4 = \left(\begin{array}{c} \overline{4\frac{1}{2}} \\ 4 \end{array}\right).$

Its optimal solution (by inspection) is

149

$$X_N^* = (x_3, x_4) = (0, 5).$$

Then, $X_B^* = (x_1, x_2) = B^{-1}b - B^{-1}NX_N^*$

$$= \begin{pmatrix} 4\frac{1}{2} \\ 4 \end{pmatrix} - \begin{pmatrix} \frac{-1}{10} & \frac{3}{10} \\ \frac{2}{5} & \frac{-1}{5} \end{pmatrix} \begin{pmatrix} 0 \\ 5 \end{pmatrix}$$

$$= \begin{pmatrix} 3 \\ 5 \end{pmatrix}$$

has $X_B^* > 0$. So, $(x_1, x_2) = (3, 5)$ provides an optimal solution to the integer problem.

Example (From [6], p. 298)

The optimal tableau

			$-x_3$	$-x_5$
(7)	M	$4\frac{2}{7}$	$\frac{5}{7}$	$\frac{3}{7}$
	x_1	$1\frac{6}{7}$	$\frac{1}{7}$	$\frac{2}{7}$
	x_2	$\frac{9}{7}$	$-\frac{2}{7}$	$\frac{3}{7}$
	x_4	$4\frac{3}{7}$	$-\frac{3}{7}$	$\frac{22}{7}$

produces the group problem

Maximize $\quad -\frac{5}{7}x_3 - \frac{3}{7}x_5$

with $\quad \begin{pmatrix} \frac{1}{7} \\ -\frac{2}{7} \\ -\frac{3}{7} \end{pmatrix} x_3 + \begin{pmatrix} \frac{2}{7} \\ \frac{3}{7} \\ \frac{22}{7} \end{pmatrix} x_5 = \begin{pmatrix} \frac{13}{7} \\ \frac{9}{7} \\ \frac{31}{7} \end{pmatrix}$

150

That is,

$$\overline{\left(\begin{array}{c} \dfrac{1}{7} \\[4pt] \dfrac{5}{7} \\[4pt] \dfrac{4}{7} \end{array}\right)} \; (x_3 + 2x_5) \; = \; \overline{\left(\begin{array}{c} \dfrac{6}{7} \\[4pt] \dfrac{2}{7} \\[4pt] \dfrac{3}{7} \end{array}\right)},$$

$$x_3, \; x_5 \geq 0, \text{ integer.}$$

Note here that G_B is cyclic of order 7. The constraint forces $x_3 + 2x_5 \equiv 6 \pmod 7$. Thus an optimal solution to GP is

$$x_3 = 0, \; x_5 = 3.$$

But in this case

$$X_B = \left(\begin{array}{c} \dfrac{13}{7} \\[4pt] \dfrac{9}{7} \\[4pt] \dfrac{31}{7} \end{array}\right) - \left(\begin{array}{cc} \dfrac{1}{7} & \dfrac{2}{7} \\[4pt] -\dfrac{2}{7} & \dfrac{3}{7} \\[4pt] -\dfrac{3}{7} & \dfrac{22}{7} \end{array}\right) \left(\begin{array}{c} 0 \\[4pt] 3 \end{array}\right)$$

$$= \left(\begin{array}{c} 1 \\ 0 \\ -5 \end{array}\right)$$

is <u>not</u> ≥ 0. Thus it is not always the case that an optimal solution to GP can be 'pulled back' to solve the original IP.

The group G_B of Gomory cuts

The group G_B is often referred to as the group of Gomory cuts. We have shown that it is a finitely generated abelian group with exponent $\leq D = \left| \det B \right|$. We sharpen our result by actually determining the order of G_B.

        As before, for the optimal basis B write $A = (B : N)$. Let Z[A] denote the abelian subgroup of Z^m generated by the columns of A. As a matter of fact $Z[A] = Z^m$ since we are operating under the assumption that A contains the m x m identity matrix as a submatrix.

        Now, consider the mapping

$$g: \quad Z[A] \longrightarrow \quad R^m/_{Z^m}$$

by $g(X) = f(B^{-1}X) = \overline{B^{-1}X}$,

        for $X \in Z[A]$.

Some observations:

1. g is a homomorphism of additive abelian groups (since f is --- recall f is the natural projection $R^m \longrightarrow R^m/_{Z^m}$).

2. Let $B_1, \ldots, B_m$ denote the columns of B, and $\{N_j\}_{j \in I}$ the columns of N. Then, if X is in Z[A] we can write

$$X = \sum_{i=1}^{m} a_i B_i + \sum_{j \in I} b_j N_j,$$

for some <u>integers</u> a_i, b_j.

So,
$$B^{-1}X = \sum_{i=1}^{m} a_i B^{-1}B_i + \sum_{j \in I} b_j B^{-1}N_j$$
and

$$g(X) = \overline{B^{-1}X} = \sum_{j \in I} b_j \overline{B^{-1}N_j},$$

since $a_i B^{-1}B_i$ is a vector with integer coordinates.

But, by definition, G_B is generated by $\overline{B^{-1}N_j}$, $j\epsilon I$. We have thus shown that g is a homomorphism onto the group G_B.

3. Further, X is in the kernel of g

⟵————⟶ g(X) is in Z^m

⟵————⟶ $\sum_{j\epsilon I} b_j B^{-1}N_j = Y = (y_i)$ is an integer vector

⟵————⟶ $\sum_{j\epsilon I} b_j N_j = BY$, for some integer vector Y

⟵————⟶ $\sum_{j\epsilon I} b_j N_j = \sum_{i=i}^{m} y_i B_i$ for some integers y_i

————⟶ X is an integral linear combination of the columns of B.

4. Let Z[B] denote the abelian group generated by the columns of B. So, the kernel of g is Z[B].

Coupling the above remarks with the Fundamental Theorem of Group Homomorphisms, we get the

(8) <u>Proposition</u>

$$G_B \approx Z[A]/Z[B] \approx Z^m/Z[B]$$

So, the finite group G_B can be thought of as the quotient of two free abelian groups of rank m.

Now, one could quote the following abstract result concerning such quotients:

If G is a free abelian group of rank m and H is a subgroup of
(9) G, also free abelian of rank m, then there exists a basis $\{g_i\}$, i = 1, ..., m, of G and positive integers c_i, i = 1, ..., m such that the collection $\{c_i g_i\}$ forms a basis of H (for a proof

of this, see [12], p. 50).

Thus the order of G/H would be given by the product $c_1 c_2 \ldots c_m$ since

$$G/H \approx \frac{Zg_1 \oplus Zg_2 \oplus \ldots \oplus Zg_m}{Zc_1 g_1 \oplus Zc_2 g_2 \oplus \ldots \oplus Zc_m g_m}$$

$$\approx Zg_1/Zc_1 g_1 \oplus \ldots \oplus Zg_m/Zc_m g_m$$

$$\approx Z/c_1 Z \oplus \ldots \oplus Z/c_m Z,$$

where $Z/c_i Z$ is the cyclic group of order c_i.

We apply this result to our situation with $G = Z^m$ and $H = Z[B]$, using the 'natural basis' $\{e_i\}$, $i = 1, \ldots, m$, for Z^m, where e_i is the unit vector with a 1 in the ith position. By definition, $\{B_i\}$ forms a basis of $Z[B]$ (B_i is the ith column of B). The matrix B can be viewed then as giving the expression of the basis elements B_i of $Z[B]$ in terms of the basis e_i of Z^m.

Standard linear algebraic arguments show that changes of bases in $Z[B]$ and Z^m correspond to multiplying B on the right and left by invertible integer matrices. Further, these 'change of basis' matrices must be unimodular -- i.e., have determinant 1 or -1 -- since their inverses are also integer matrices. (The arguments needed to verify this in detail are identical to those used in the crystallography chapter to discuss change of lattice bases.)

Thus, the abstract result (9) implies that there exist unimodular matrices U and V with

$$V B U = \begin{pmatrix} c_1 & & & & \bigcirc \\ & c_2 & & & \\ & & c_3 & & \\ & & & \ddots & \\ \bigcirc & & & & c_m \end{pmatrix}.$$

But this can now be used to derive an expression for the order of G_B.

We have order of $G_B = c_1 c_2 \cdots c_m = \left| \det \begin{pmatrix} c_1 & & & \bigcirc \\ & c_2 & & \\ & & \ddots & \\ \bigcirc & & & c_m \end{pmatrix} \right|$

$$= \left| \det (VBU) \right|$$

$$= \left| \det B \right|, \text{ using the fact that } \left| \det V \right| = \left| \det U \right| = 1.$$

We have shown

Proposition   The order of G_B is $D = \left| \det B \right|$.

Example

        The group G_B of Gomory cuts in Example (6) has order

$$\left| \det \begin{pmatrix} 2 & 3 \\ 4 & 1 \end{pmatrix} \right| = 10.$$

Remark

        The abstract result (9) is actually equivalent to asserting that any non-singular integer matrix can be diagonalized over the ring of integers using elementary row and column operations. This is sometimes referred to as reducing the matrix to Smith Normal Form. Constructive algorithms effecting this reduction can be found in [3] and [4, p. 268]. Use of these would permit bypassing the existence result (9).

A bound on $\sum x_i$

Since the order of $\bar{\alpha}_i$ in G_B is less than or equal to D, an optimal solution $X = \{x_i\}$ to the group problem GP will certainly have $0 \leqslant x_i \leqslant D - 1$. Interestingly, a tighter a priori upper bound can be given.

Proposition   An optimal solution $X_N = (x_j)_{j \in I}$ to the group problem

$$\text{Maximize} \quad (C_N - C_B \ B^{-1}N) \ X_N$$

$$\text{subject to} \quad \sum_{j \in I} \bar{\alpha}_j \ x_j = \bar{\alpha}_0 \ , \ x_j \geqslant 0, \text{ integer}$$

can be found with $\sum_{j \in I} x_j \leqslant D - 1$.

Proof:

For ease of exposition we let the index set I consist of $\{1, 2, \ldots, r\}$ and drop the bars on the $\bar{\alpha}_j$. Thus the group constraint is written as

$$(*) \qquad \sum_{j=1}^{r} x_j \alpha_j = \alpha_0 \ .$$

Now, suppose $X = \{x_j\}$ is an optimal solution to GP chosen with $\sum_{j=1}^{r} x_j$ minimal.

Suppose that $\sum_{j=1}^{r} x_j > D$. Consider the list of all the group elements α_j appearing on the left side of $(*)$, with α_j appearing x_j times (throw in also 0):

$$0, \ \underbrace{\alpha_1, \ \alpha_1, \ \ldots, \ \alpha_1}_{x_1 \text{ times}}, \ \underbrace{\alpha_2, \ \ldots, \ \alpha_2}_{x_2 \text{ times}}, \ \ldots$$

Form the sequence S_i of partial sums of this list:

$$S_0 = 0$$

$$S_1 = \alpha_1$$

$$S_2 = \alpha_1 + \alpha_1$$

$$S_3 = \alpha_1 + \alpha_1 + \alpha_1$$

$$\cdots$$

$$\cdots$$

$$S_{x_1} = x_1\alpha_1$$

$$S_{x_1+1} = x_1\alpha_1 + \alpha_2$$

$$\cdots$$

$$\cdots$$

$$S_{\sum x_i} = \sum x_i\alpha_i$$

Note that the S_i's are elements of the group G_B and, by assumption, there are at least $D + 1$ of them. Since $\left|G_B\right| = D$, some two of these must be the same. Thus we are assured that

$$S_i = S_j \text{ for some i and j, } 0 < i < j.$$

Then $S_j - S_i = 0$ and

$$\sum x_i\alpha_i = \sum x_i\alpha_i - (S_j - S_i).$$

But, $\sum x_i\alpha_i - (S_j - S_i)$ can be written in the form $\sum y_i\alpha_i$ where

157

$0 < y_i < x_i$, y_i integer with at least one y_i strictly less than the corresponding x_i.

So, the vector $Y = (y_i)$ provides a feasible solution to GP with

$$(C_N - C_B B^{-1} N) \ X < (C_N - C_B B^{-1} N) \ Y.$$

But $\sum_i y_i < \sum_i x_i$, thus contradicting the fact that X was optimal with $\sum_i x_i$ minimal.

Example

In Example (6) we are guaranteed that an optimal solution to the group problem will satisfy $x_3 + x_4 < 10-1 = 9$.

Concluding Remarks

1. Our discussion can be summarized as follows:

   1) To solve an integer programming problem of type (5), use the simplex algorithm to find an optimal basis B for the problem considered as a linear program.

   2) Set-up and solve the group problem GP.

   3) If $X_B = B^{-1}b - B^{-1}NX_N > 0$, then X_B provides an optimal solution to the original problem.

2. The finiteness of the group G_B along with the bound $\sum_j x_j < D - 1$ certainly guarantee that the group problem can be solved by finite enumeration. The difficulty is that for large values of D the computations become prohibitively time consuming and expensive. In [7] Gomory suggests that a technique called dynamic programming be used in solving GP (cf. also [1], [4]). This technique seems to be particularly useful when G_B is cyclic (cf. Exercise 10). The

examples and exercises in this section are deliberately chosen to be amenable to solution by inspection.

3. It is possible to give conditions under which the solution to GP can always be transferred to a solution X_B of the original problem. One such is:

> Let $l = $ maximum of the absolute values of the entries in $B^{-1}N$.
>
> Then, if $B^{-1}b \geq (D-1)l$, it follows that
>
> $X_B \geq 0$. (c.f. [7], [10]).

The problem is that this condition is not necessary and is too restrictive in practice.

4. The Fundamental Theorem of Finite Abelian Groups allows the decomposition of G_B in the form

$$G_B \simeq Z_{q_1} \oplus Z_{q_2} \oplus \ldots \oplus Z_{q_k}$$

where q_i, the order of the cyclic group Z_{q_i}, , divides q_{i+1}, $i=1, \ldots, k-1$.

Thus elements of G_B can be represented by ordered k-tuples of the form

$$(g_1, g_2, \ldots, g_k), \quad 0 \leq g_i \leq q_i - 1.$$

Addition is modular and done componentwise using the modulus q_i in the ith component. This representation is useful in designing and coding computer algorithms to solve the group problem GP (cf. [8]).

5. We close with an <u>Example</u>:

$$\begin{aligned}
\text{Maximize} \quad & 2x_1 + 3x_2 \\
\text{subject to} \quad & x_1 + 2x_2 \leq 12 \\
& 4x_1 + x_2 \leq 32
\end{aligned}$$

An optimal LP tableau is

	$-x_3$	$-x_4$
$21\frac{5}{7}$	$\frac{10}{7}$	$\frac{1}{7}$
x_1 $\quad \frac{52}{7}$	$\frac{-1}{7}$	$\frac{2}{7}$
x_2 $\quad \frac{16}{7}$	$\frac{4}{7}$	$\frac{-1}{7}$

and reveals that

$$B = \begin{pmatrix} 1 & 2 \\ 4 & 1 \end{pmatrix} \qquad\qquad B^{-1}N = \begin{pmatrix} -\frac{1}{7} & \frac{2}{7} \\ \frac{4}{7} & -\frac{1}{7} \end{pmatrix}$$

$$B^{-1}b = \begin{pmatrix} \frac{52}{7} \\ \frac{16}{7} \end{pmatrix} \qquad\qquad C_N - C_B B^{-1}N = (-\frac{10}{7}, -\frac{1}{7})$$

The corresponding group problem is

Maximize $\qquad\qquad -\frac{10}{7} x_3 - \frac{1}{7} x_4$

subject to

$$x_3 \begin{pmatrix} \overline{-\frac{1}{7}} \\ \overline{\frac{4}{7}} \end{pmatrix} + x_4 \begin{pmatrix} \overline{\frac{2}{7}} \\ \overline{-\frac{1}{7}} \end{pmatrix} = \begin{pmatrix} \overline{\frac{52}{7}} \\ \overline{\frac{16}{7}} \end{pmatrix}$$

$$x_3, x_4 > 0.$$

Note that $\left| G_B \right| = D = \left| \det B \right| = 7$, and being of prime order is cyclic.

Taking $\overline{\left(\begin{array}{c} \frac{2}{7} \\ -\frac{1}{7} \end{array}\right)} = \overline{\left(\begin{array}{c} \frac{2}{7} \\ \frac{6}{7} \end{array}\right)}$ as a generator, the GP constraint becomes

$$(3x_3 + x_4) \overline{\left(\begin{array}{c} \frac{2}{7} \\ \frac{6}{7} \end{array}\right)} = \overline{\left(\begin{array}{c} \frac{3}{7} \\ \frac{2}{7} \end{array}\right)}.$$

Hence, we must have

$$3x_3 + x_4 \equiv 5 \ (\text{mod } 7)$$

and $0 \leqslant 3x_3 + x_4 \leqslant D - 1 = 6.$

The only possibilities are

$$x_3 = 0, \ x_4 = 5 \ \text{or} \ x_3 = 1, \ x_4 = 2.$$

Clearly $X_N = \left(\begin{array}{c} 0 \\ 5 \end{array}\right)$ is optimal for the GP.

Then

$$X_B = B^{-1}b - B^{-1}NX_N$$

$$= \left(\begin{array}{c} \frac{52}{7} \\ \frac{16}{7} \end{array}\right) - \frac{1}{7} \left(\begin{array}{cc} -1 & 2 \\ 4 & -1 \end{array}\right) \left(\begin{array}{c} 0 \\ 5 \end{array}\right)$$

$$= \left(\begin{array}{c} 6 \\ 3 \end{array}\right)$$

Since $X_B \geqslant 0$ we are guaranteed that $x_1 = 6$ and $x_2 = 3$ provides an optimal solution to the IP.

References

[1] M. L. Balinski   "Integer Programming:
                                         Methods, Uses, Computation,"
                                         Management Science 12
                                         (1965), p. 253-313.

[2] M. Bellmore and G. L. Nemhauser   "The Traveling Salesman
                                         Problem: A Survey,"
                                         Operations Research 16
                                         (1968), p. 538-558.

[3] Gordon H. Bradley   "Algorithms for Hermite and
                                         Smith Normal Matrices and
                                         Linear Diophantine
                                         Equations," Mathematics of
                                         Computation, Vol. 25, No.
                                         116, October 1971, p. 897-
                                         907.

[4] R. S. Garfinkel and G. L. Nemhauser Integer Programming, John
                                         Wiley & Sons, New York,
                                         1972.

[5] Saul I. Gass   Linear Programming (4th
                                         Edition), McGraw-Hill, New
                                         York, 1975.

[6] Ralph E. Gomory   "An Algorithm for Integer
                                         Solutions to Linear
                                         Programs" in Recent Advances
                                         in Mathematical Programming,
                                         R. L. Graves & P. Wolfe
                                         eds., McGraw-Hill, 1963.

[7] _____   "On the Relation Between
                                         Integer and Noninteger
                                         Solutions to Linear
                                         Programs," Proceedings
                                         National Academy of Sciences
                                         53 (1965), p. 260-265.

[8] G. A. Gorry, W. D. Northup, J. Shapiro "Computational Experience
                                         with a Group Theoretic
                                         Integer Programming
                                         Algorithm," Mathematical
                                         Programming 4 (1973) p. 171-
                                         192.

[9] B. Kolman and R. Beck   Elementary Linear
                                         Programming with
                                         Applications, Academic
                                         Press, 1980.

[10] K. O. Kortanek and R. Jeroslow   "An Exposition on the
                                        Constructive Decomposition
                                        of the Group of Gomory Cuts
                                        and Gomory's Round-off
                                        Algorithm," Cahiers du
                                        Centre d'Etudes de Recherche
                                        Operationnelle, Vol. 13,
                                        1971, p. 63-84.

[11] R. Von Randow, editor   Integer Programming and
                                       Related Areas: A Classified
                                       Bibliography 1978-1981,
                                       Lecture Notes in Economics
                                       and Mathematical Systems
                                       197, Springer Verlag, 1982.

[12] Eugene Schenkman   Group Theory, Van Nostrand,
                                        Princeton, 1965.

Exercises

Solve the following integer programs using the group reduction
techniques of this section. In each instance find the order of the
group G_B.

1. Maximize $x_1 + x_2$

   subject to $x_1 \leq 1$

                    $x_2 \leq 1$

                    $x_1, x_2 \geq 0$, integer.

2. Maximize $-x_1 + x_2$

   subject to $x_1 + 3x_2 \leq 67$

                    $2x_1 + x_2 \leq 64$

                    $x_1, x_2 \geq 0$, integer.

3. Maximize $6x_1 + 5x_2$

   subject to $x_1 + x_2 \leq 4$

$$5x_1 + 3x_2 \leqslant 15$$

$$x_1, x_2 \geqslant 0, \text{ integer.}$$

4. Maximize $\qquad x_1 + 2x_2$

subject to $\qquad x_1 + 2x_2 \leqslant 8$

$$2x_1 \leqslant 7$$

$$-2x_1 + 4x_2 \leqslant 9$$

$$x_1, x_2 \geqslant 0, \text{ integer.}$$

5. Consider the following optimal tableau.

		$-x_3$	$-x_5$
	$4\frac{2}{7}$	$\frac{5}{7}$	$\frac{3}{7}$
x_1	$1\frac{6}{7}$	$\frac{1}{7}$	$\frac{2}{7}$
x_2	$\frac{9}{7}$	$-\frac{2}{7}$	$\frac{3}{7}$
x_4	$4\frac{3}{7}$	$-\frac{3}{7}$	$\frac{22}{7}$

1) Find the optimal solution X_N to the corresponding group problem.

2) Show that in this case it is not true that $X_B = B^{-1}b - B^{-1}NX_N$ satisfies $X_B \geqslant 0$.

6. Find the actual integer optimum in Example (7).

7. What is the structure of the group G_B in Example (6)?

8. Prove: If the matrix B corresponding to an optimal basis is unimodular, then the corresponding optimal LP solution has integer coordinates.

9. Show that if $G_B = \{0\}$, then an optimal LP solution corresponding to the basis B has integer coordinates.

10. Let $\left| G_B \right| = D$. Suppose some element $\bar{\alpha}_j$ in G_B (considered as a column vector with fractional entries) contains an entry of the form $\frac{r}{D}$, with r relatively prime to D. Prove that then G_B is cyclic.

11. Suppose the requirement that A contain the m x m identity matrix as a submatrix is dropped and we assume only that A has rank m.
    1) Mimic the proof of Proposition (8) to show that $G_B \approx Z[A]/Z[B]$.
    2) Deduce that the order of G_B divides $D = \left| \det B \right|$.
       (Hint: $Z[A]/Z[B]$ is a subgroup of $Z^m/Z[B]$).

10   Group Theory and Counting

The question of counting the number of ways something can be done is one that arises often in applications. The branch of mathematics dealing with the art of sophisticated counting is called combinatorics and is one which has contributed not only to established disciplines like probability, statistics and graph theory, but has more recently had an impact on emerging areas of theoretical computer science, most notably algorithm design and coding. Symmetry frequently enters into problems of counting, so it should not be surprising that group theory might at times be brought to bear on such problems. In this chapter we develop one elementary group theoretic result in counting, called Burnside's Lemma. Extensions of this result were used by the mathematician George Polya to solve enumeration problems arising in chemistry. To set the scene we begin with a concrete example, which we then generalize; having solved the general problem we return to several applications.

Motivating Example

Seating at an awards banquet attended by both faculty and students is at rectangularly shaped tables. A given table might seat any combination of students and faculty members. For example,

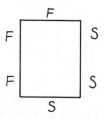

where F denotes a faculty member and S a student, is a possible arrangement. In thinking about the number of ways seating of this type can be arranged, organizers of the affair realize that they are concerned only about "relative seating." That is, the above distribution is regarded the _same_ as the seatings

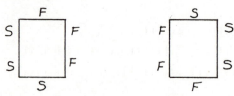

All three have two F's sitting opposite two S's and an F and S facing each other at opposite heads of the table.

Certainly, <u>without</u> thinking of certain arrangements as being identical, there would be 2^6 = 64 possible seatings (there are six places to be filled, with two possibilities for each place). But, by making identifications of the type above, the number of seatings considered different by the organizers should be considerably smaller.

The question posed then is: What is the number of seating arrangements considered different by the organizers? The answer could reasonably be arrived at by hand enumeration, but note that the problem would be a bit more complicated if, say, administrators were also admitted or if the shape of the table were made more exotic. Symmetries of the table cause certain seatings to be identified and we will use this relatively simple problem to motivate a more general discussion of the effect of symmetries on counting. Where symmetry is present groups are not far behind, and we proceed to attach an appropriate group to this problem.

If we think of a table as being centered about horizontal and vertical axes,

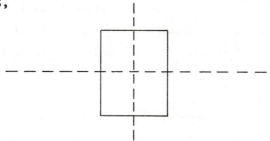

then a more precise way of saying that two seatings are considered identical is to say that one of them is obtained from the other by some combination of reflections about these axes.

Let α = reflection about the x axis

and β = reflection about the y axis.

Then the set G = $\{ 1, \alpha, \beta, \alpha\beta\}$ forms a group, where the operation is composition of reflections ($\alpha\beta$ means first reflect about the y axis, then the x axis). Thus G is abelian, with each element of G having order 2 (G is often referred to as Klein's four-group).

Now, let X denote the set of all <u>non-identified</u> 64 seating arrangements. So,

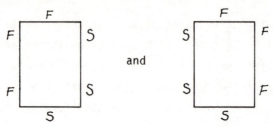

and

are distinct elements of X. Again, $|X| = 2^6$.

For a seating arrangement x in X and element g of the group G, we use the notation gx to denote the seating arrangement resulting from x by applying the symmetry g. For example, if x is

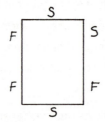

and g = αβ, then gx is

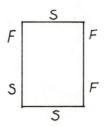

The point is that for a seating x, it is exactly all seatings of the form gx, as g varies through the elements of the group G, that the organizers consider the same as x.

This can be formalized by defining an equivalence relation in the set X. We introduce a relation ∼ in X by

(1)   x ∼ y   if and only if there exists a g ε G with gx = y.

The relation ∼ is an equivalence relation (cf. Exercise 2) and all seating arrangements in a fixed equivalence class are one and the same for the organizers. The question we are posing in this problem can then

168

be rephrased as: Find the <u>number of equivalence classes</u> X possesses under the relation ~ . We will compute this number shortly, but pause first to remark that the situation presented by this problem can easily be generalized.

The General Situation

A general version of the above problem would envision a set X of yet unidentified objects and a group G that is used to express when two objects in X are to be considered the same. It does no harm in the following to imagine X as playing the role of the 64 original possible seatings and G as the group of reflections presented above.

More specifically, suppose X is a finite set.

<u>Def.</u>   The finite group G is a <u>group of symmetries of X</u> if and only if

1) to each $g \in G$ and $x \in X$ we can attach a unique element gx in X,

2) 1x = x, for all x in X and

3) for all g, h in G it is the case that g(hx) = (gh)x.

Remarks

1. In the seating arrangement problem, G = $\{1, \alpha, \beta, \alpha\beta\}$ is a group of symmetries of X.

2. Suppose a triangle could have either a yellow or blue balloon attached to any one of its three vertices. Let X consist of all possible triangles that can result in this way.

So,   and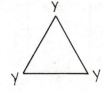

are two of the eight elements in X.

Let R designate the operation of rotating a triangle counter-clockwise through 120 degrees. Then $R^3 = 1$ (rotating three times is the same as doing nothing) and G = $\{1, R, R^2\}$ is a group of symmetries of the set X. Here, <u>e.g.</u>, by R^2x is meant the triangle resulting by rotating x through 240 degrees.

3. Note that, if h ε G and x ε X , then by 1) hx is an element of X.
    Thus, again by 1), g(hx) makes sense. Also, if g is an element of
    the group G, then gh is in G, and so (gh)x is defined. Requirement
    3) in the definition forces the two elements g(hx) and (gh)x to be
    the same.

    Exactly as in the motivating problem, if G is a group of
    symmetries of X, G can be used to define a relation on X by

    x ~ y   if there exists a g in G with y = gx.

This yields an equivalence relation in X and hence it makes sense to
speak of the equivalence class of an element of X, the notation for
which follows.

Def.   For x ε X,
        cl(x) = equivalence class of x under ~ .

Examples
1. In the seating arrangement problem, if x is

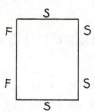

then cl(x) consists of

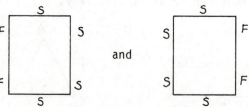

and

2. In the triangles with balloons example, if

x is

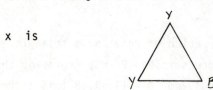

170

then cl(x) consists of

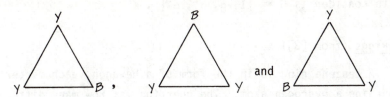

and

The general problem we pose is: given a group of symmetries of the set X, what is the number of equivalence classes induced by this group in X? Before analyzing this general question, we list some concrete problems that can be interpreted using groups of symmetries.

(Polya) Necktie Problem

A straightforward necktie has 10 bands, each of which can be colored by any of 4 colors.

How many distinct neckties of such type are there? Here it is appropriate to note that it makes no difference which end we view the necktie from. If we let α designate the process of reflecting a necktie about its center, then the colored necktie x is viewed as being the same as the necktie αx. If X is the set of all possible colorings of neckties, the $|X| = 4^{10}$. A group of symmetries of X appropriate to this problem is G = $\{1, \alpha\}$.

Color Wheels

A wheel has 4 compartments, each of which can be colored by any one of 3 colors.

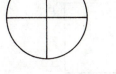

Find the number of such wheels. Here, X is the set of all possible colorings of the wheel with $|X| = 3^4$.

Letting α denote a rotation through 90°, the group of symmetries to consider is G = $\{1, \alpha, \alpha^2, \alpha^3\}$.

Benzene Rings (from [3])

A benzene ring is in the form of a hexagon, each vertex of which contains a hydrogen atom. The question is, How many different compounds can be obtained by substituting chlorine atoms for the hydrogen atoms in the ring? One such possibility is indicated here -- empty vertices contain hydrogen atoms.

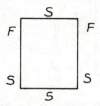

Here, the relevant group G would be the group of symmetries of the hexagon --- namely, the dihedral group D_6. The set X would consist of all possible placements of chlorine atoms without regard to the symmetry of the molecule. So, here, $|X| = 2^6 = 64$.

Some Concepts and Notation

For the remainder of this chapter we will always assume that X is a finite set and that G is a group of symmetries of X. To pursue our aim of counting the number of equivalence classes in X, we introduce several concepts.

Def. For x ε X, <u>the point stabilizer of x,</u> H_x, is that subgroup of G given by $H_x = \{g \epsilon G | gx = x\}$.

Examples and Remarks

1. in the motivating problem, if x is

the $H_x = \{1, \beta\}$.

2. The reason for the terminology is that H_x consists of those elements of the symmetry group that fix or stabilize x.

3. In the color wheel problem, if x is

where Bu denotes blue and Y denotes yellow, then $H_x = \{1, \alpha^2\}$.

The point stabilizer of x gives us a way of counting the number of elements in the equivalence class of x. This is the content of the

<u>Proposition</u> | cl(x) | = index of H_x in G

               $= [G:H_x]$.

<u>Example</u>   The equivalence class of

$$x =$$

contains two elements, while $H_x = \{1, \alpha\}$.

Since |G| = 4, the statement of the Proposition clearly holds in this case.

<u>Proof of the Proposition</u>:

Writing H instead of H_x, we note that by definition [G:H] is equal to the number of left (or right) cosets of H in G. We let G/H denote the set of left cosets of H in G.

The Proposition will be proved if we can show that there are as many elements in G/H as there are elements in cl(x) and we proceed to do this by defining a function between these two sets which will turn out to be a 1-1 correspondence.

Define the function f: G/H ----→ cl(x) by f(gH) = gx, where g∊G.

173

1) Since coset representatives are not unique, the first item of business is to show that this function is well-defined: _i.e._, we must show that if gH = rH, it follows that gx = rx. So, suppose gH = rH. This means that $g^{-1}r \varepsilon H$ and, since H is the point stabilizer of x, we get that $(g^{-1}r)x = x$. Multiplying both sides of this relation on the left by g, we get the desired result rx = gx.

2) f is onto (or surjective) since, if $y \varepsilon cl(x)$, then y = gx for some g in G -- here we are using the definition of the equivalence relation induced by G. So, f(gH) = gx = y and the coset gH is mapped onto y by the function f.

3) We show that f is 1-1 by showing that if two cosets have identical images, they must have been equal cosets at the start. So, suppose that f(gh) = f(rH). Then gx = rx or $(r^{-1}g) x = x$. But this means that $r^{-1}g \varepsilon H_x = H$. The definition of coset equality then says gH = rH. This ends the proof.

An interesting sidelight of the above result is the determination of the orders of the rotation groups of the regular three dimensional solids. For example, let G be the group of rotations of the cube. Letting X be the set of faces of the cube, then G is certainly a group of symmetries of X ($|X| = 6$).

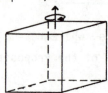

Let x denote the topmost face of the cube. Since any face can be rotated into any other face, it follows that $|cl(x)| = 6$. The point stabilizer of x, H_x, consists of all rotations that leave the topmost face fixed -- these are precisely the rotations about the axis passing through the center of x. Including the identity there are 4 such rotations, so $|H_x| = 4$. The Proposition says that

$$|cl(x)| = [\ G: H_x] = \frac{|G|}{|H_x|}\ .$$

So,

$$|G| = |cl(x)| \cdot |H_x|\ .$$

174

In our situation this gives
$$|G| = 6 \cdot 4 = 24$$
and thus the group of rotations of the cube has 24 elements. Similar arguments may be given to determine the orders of the rotation groups of the other regular solids: the tetrahedron, the octahedron, the dodecahedron and the icosahedron.

Another notion we introduce is that of the number of items in X that an element of the symmetry group G fixes.

Def.   For $g \in G$,
$$F(g) = \underline{\text{number}} \text{ of elements of X fixed by } g$$
$$= |\{x \in X \mid gx = x\}| \ .$$

Examples
1. In the motivating problem,

$$F(1) = 2^6 \qquad\qquad F(\beta) = 2^4$$
$$F(\alpha) = 2^3 \qquad\qquad F(\alpha\beta)= 2^3 \ .$$

2. In the color wheel problem with 3 colors,
$$F(\alpha) = 3, \quad \text{while } F(\alpha^2) = 3^2 \ .$$
3. It is always the case that $F(1) = |X|$.

Our basic question was: Find the number of equivalence classes in X induced by G. The answer to that question is contained in a result which dates back to the turn of the 20th century often credited to the English mathematician W. Burnside.

Burnside's Lemma   Let G be a group of symmetries of the set X. If N is the number of equivalence classes in X induced by G, then

$$N = \frac{1}{|G|} \sum_{g \in G} F(g) \ .$$

Example   (the motivating problem)

$$N = \frac{1}{4} (F(1) + F(\alpha) + F(\beta) + F(\alpha\beta))$$

$$= \frac{1}{4} (2^6 + 2^3 + 2^4 + 2^3) = 24 \ .$$

175

So, the organizers can find 24 distinct types of faculty-student seating arrangements.

Proof of Burnside's Lemma:

For $g \in G$, think of the number $F(g)$ as a sum of ones -- $F(g) = 1+1+\cdots+1$ -- where a one is contributed for each element x of X that g fixes. So, $S = \sum_{g \in G} F(g)$ is also a larger sum of ones.

Now imagine each of the ones in the larger sum S as being identified or tagged by the element x in X it arises from. We now ask, Given a fixed x in X, how many ones in S carry the identification tag "x"? The answer is precisely $|H_x|$, since this is the number of elements in the symmetry group G fixing x. It is a fact that each element in cl(x), the equivalence class of x, also contributes exactly the same number of ones -- we deal with the proof of this fact separately in the "Sub-proof" below.

Accepting this, it follows that the elements in cl(x) contribute a grand total of $|cl(x)||H_x|$ ones to the sum S. But, by a previous Proposition,

$$|cl(x)||H_x| = \frac{|G|}{|H_x|} |H_x| = |G| .$$

So the elements of each equivalence class cl(x) account for $|G|$ ones. But, since X is the disjoint union of its N equivalence classes, it follows that the total number of ones in S is given by $|G| \cdot N$. That is, $\sum_{g \in G} F(g) = |G| \cdot N$, or no. of equivalence classes = $N = \frac{1}{|G|} \sum_{g \in G} F(g)$.

This is our desired result.

Examples

1. ### Necktie Problem

With 4 colors and 10 bands, we have $|X| = 4^{10}$. Also, $G = \{1, \alpha\}$ where $\alpha^2 = 1$. Then, $F(1) = 4^{10}$ and $F(\alpha) = 4^5$, since for a necktie to be fixed by α, the colors in the first five bands must be reproduced in the last five bands. Thus, by Burnside's Lemma,

176

$$\text{no. of neckties} = \frac{1}{2}(4^{10} + 4^5) .$$

2. Color Wheel

Here, $|X| = 3^4$ and $|G| = 4$. Computing we get $F(1) = 3^4$, $F(\alpha) = 3$, $F(\alpha^2) = 3^2$ and $F(\alpha^3) = 3$. So,

$$\text{no. of wheels} = \frac{1}{4}(3^4 + 3^2 + 2 \cdot 3)$$

3. Benzene Rings

Here, $G = D_6$, the group of symmetries of the hexagon. The 12 elements of the dihedral group D_6 are

$$1, R, R^2, R^3, R^4, R^5, F, FR, FR^2, FR^3, FR^4, FR^5$$

where R is a rotation through 60^0 and F is a flip through a fixed axis of the hexagon. (For more information on dihedral groups consult the chapter on crystallography).

A lengthy computation of the number of items of X fixed by each of these group elements yields that

$$\text{no. of compounds} = \frac{1}{12}(2^6 + 2^2 + 2^3 + 2^5 + 3 \cdot 2^4)$$

$$= 13 .$$

'Sub-proof' for use in Burnside's Lemma

Here we must show that if x and y are equivalent then $|H_x| = |H_y|$.

So, suppose $y \sim x$. Then there exists a $g \varepsilon G$ with $y = gx$. We then claim that $H_y = gH_xg^{-1}$ and we verify this by showing that each is a subset of the other.

Let $r \varepsilon gH_xg^{-1}$. Then $r = gsg^{-1}$ for some s in H_x. So,

$$ry = (gsg^{-1}) gx = g(sx)$$
$$= gx , \text{ since } s \varepsilon H_x$$
$$= y.$$

That is, $r \varepsilon H_y$.

Conversely, suppose $r \varepsilon H_y$. Then $ry = y$ and so, $r(gx) = gx$. Multiplying on the left by g^{-1} yields $(g^{-1}rg) x = x$, which says that $g^{-1}rg \varepsilon H_x$. Put another way, $r \varepsilon g H_x g^{-1}$.

Since $g H_x g^{-1} = \{ghg^{-1} | h \varepsilon H_x\}$, the sets $g H_x g^{-1}$ and H_x are in 1-1 correspondence. It follows that $|g H_x g^{-1}| = |H_x|$. That is $|H_y| = |H_x|$ and the proof is complete.

The Color Wheel problem can be readily generalized to a wheel with n compartments where each compartment can be colored by any one of q colors. We include a discussion of this generalization since it simultaneously involves invoking some interesting elementary group theory.

In this situation the appropriate group of symmetries is the cyclic group of order n, $G = Z_n$, generated by a rotation α through an angle of $2\pi/n$. The following is an outline of a derivation of the number of distinct such wheels where we ask the interested reader to supply some of the intermediate reasoning.

The number of wheels fixed by an element of G depends on the order of the element. If $g \varepsilon G$ has order k, then g fixes $q^{n/k}$ wheels. Note that k must always divide n. Given that k divides n, how many elements of $G = Z_n$ have order k? The answer is $\phi(k)$. This is so since there is at least one -- namely, $\beta = \alpha^{n/k}$ -- and all other such elements must be in the cyclic subgroup of G generated by β. The powers β^i that have the same order as β are precisely those for which $(i,k) = 1$, where $1 \leq i \leq k$. By the definition of the Euler-phi-function, there are $\phi(k)$ such elements in G.

Thus, in $\sum_{g \varepsilon G} F(g)$, the term $q^{n/k}$ occurs $\phi(k)$ times. By Burnside's Lemma, the number of distinct wheels is

$$\frac{1}{n} (\sum_{k|n} \phi(k) q^{n/k}),$$

where the sum in the brackets ranges over all possible divisors k of n. Note that we have also supplied a roundabout proof of the fact that, for any number q, the bracketed term must always be a multiple of n.

178

An Application to Group Theory

Counting arguments can often be used to great advantage in analyzing groups themselves. The number of elements in a finite group can strongly influence the structure of the group. A fine example of this is the result that, for p a prime, any group of order p^2 must be abelian. We include a derivation of this result since it can be arrived at fairly quickly using the tools of this chapter.

Suppose G is a finite group. The set X we choose for which G is a group of symmetries is G itself. That is, we let X = G and define the expression gx, for $g \varepsilon G$, $x \varepsilon X$, by

$$gx = g^{-1}xg \ .$$

This requires a bit of explanation. The product on the right is a product of three elements in the group G (the conjugate of x by g) and is hence an element of X = G. The expression gx on the left is <u>not</u> to be interpreted as a group product of the two elements, but rather is meant only to designate the element of X that x is sent to by the action of g. We leave it as an exercise to show that under the above definition G is indeed a group of symmetries of X.

Now cl(x) consists of those elements equivalent to x. In this instance, cl(x) = $\{g^{-1}x \ g | g \varepsilon G\}$ and is commonly called the conjugacy class of x.

A conjugacy class consists of only one element precisely when cl(x) = $\{x\}$ and this will happen when $g^{-1}xg = x$, for all g in G. So, cl(x) = $\{x\}$ exactly when xg = gx for all g in G -- namely, when x is in the center of G, Z(G). But the set X = G is the disjoint union of its equivalence (conjugacy) classes. So one way to compute the order of G is to compute the number of elements in each equivalence class and sum up these numbers. The order of the center of G, $|Z(G)|$, accounts for the number of equivalence classes containing one element. Keeping this in mind, we get

$$|G| = |Z(G)| + \sum |cl(x)|,$$

where the sum on the right involves one x from each equivalence class containing more than one element.

179

H_x, the stabilizer of x, consists of those g∊G that fix x. This translates here to consisting of those g for which $g^{-1}xg = x$ -- i.e. for which xg = gx. So, H_x consists of those elements that commute with x (this subgroup is called the centralizer of x). A Proposition above says that $|cl(x)| = [G:H_x]$. So, if $|cl(x)| > 1$, this must force $|H_x| < |G|$. Our previous identity can then be re-written in the form

$$|G| = |Z(G)| + \sum [G:H_x]$$

where the sum on the right is subject to the same restrictions as before.

Now we specialize to the case where $|G| = p^2$, p a prime. Since all the terms in $\sum [G:H_x]$ have $|H_x| < |G|$, it follows that for each of them $[G:H_x] = p$ or p^2. In particular, the prime p divides $\sum [G:H_x]$. Since p divides $|G|$ by hypothesis, we get as a consequence that p must divide $|Z(G)|$. By Lagrange's Theorem the only possibilities for $|Z(G)|$ are then p or p^2. We proceed to so show that $|Z(G)| = p$ is impossible.

Suppose $|Z(G)| = p$. Then Z(G) is a proper subgroup of G and there is some element, say a, which is in G, but not in Z(G). Let H_a denote the centralizer of a -- i.e. the set of those elements in G which commute with a. Certainly, Z(G) is a subgroup of H_a and a proper sub-group at that, since a is in H_a but not in Z(G). So, H_a must have order larger than p. The only out then is for H_a to have order p^2. This says that $H_a = G$ and so every element in G commutes with a. But this contra-dicts the assumption that a is not in Z(G).

Thus $|Z(G)| = p^2$ and Z(G) = G. That is, G is abelian and we have our desired result. One must admit that there is something satis-fying about being able to declare in advance that, say, any group con-sisting of 49 elements must of necessity be commutative.

Concluding Remarks

The use of Burnside's Lemma requires the computation of F(g) for each g in the symmetry group G and this can at times be a tedious task. In his paper [5] on applying group theory and combinatorics to

enumeration problems in chemistry, George Polya developed a more elegant
and feasible way of doing these computations. For an elementary exposition of Polya's theory see [8] or [9], while [1] and [3] afford a more
sophisticated look, along with applications to counting problems in
graph theory. [4] contains an application of the techniques developed
in this chapter to counting the number of essentially different ways of
placing non-attacking rooks on a chessboard. The question of whether
Burnside's Lemma is really due to Burnside is addressed in the note
[10].

References

[1] N. G. De Bruijn   "Polya's Theory of Counting" in
                                    Applied Combinatorial Mathematics,
                                    E. F. Beckenbach ed., John Wiley
                                    and Sons, 1964.

[2] Solomon W. Golomb   "A Mathematical Theory of Discrete
                                    Classification," Information
                                    Theory, Fourth London Symposium,
                                    Butterworths, London, 1961.

[3] F. Harary, E. M. Palmer and   "Polya's Contributions to Chemical
     R.W. Robinson   Enumeration" in Chemical
                                    Applications of Graph Theory, A. T.
                                    Balaban ed., Academic Press,
                                    London, 1976.

[4] D. F. Holt   "Rooks Inviolate," Mathematical
                                    Gazette, Vol. 58, 1974, pp. 131-
                                    134.

[5] G. Polya   "Kombinatorishe Anzahlbestimmungen
                                    für Gruppen, Graphen und chemische
                                    Verbindungen,"Acta Math. 68, 1938,
                                    pp. 145-254.

[6] John Riordan   "The Combinatorial Significance of
                                    a Theorem of Polya," J. Soc.
                                    Indust. Appl. Math., Vol. 5, No. 4,
                                    December 1957, pp. 225 -237.

[7] L. W. Shapiro   "Finite Groups Acting on Sets with
                                    Applications," Mathematics
                                    Magazine, May-June 1973, pp. 136-
                                    147.

[8]   Alan Tucker   _Applied Combinatorics_, John Wiley & Sons, 1980.

[9]   _____   "Polya's Enumeration Formula by Example," Mathematics Magazine, Nov.-Dec. 1974, pp. 248-256.

[10] E. M. Wright   "Burnside's Lemma: A Historical Note," Journal of Combinatorial Theory, Series B 30, 1981, pp. 89-90.

Exercises

1. Show that the group G of the motivating example is isomorphic to $Z_2 \times Z_2$.

2. Show that the relation defined in (1) is indeed an equivalence relation.

3. Compute the point stabilizers for the following elements of the set X in the motivating example.

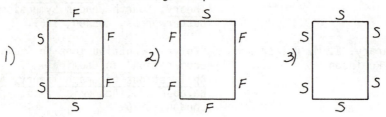

4. For $x \epsilon X$, show that H_x is a subgroup of G.

5. Four spheres are fixed onto the corners of a square by four rods. We wish to paint each of the spheres either red, white or blue.

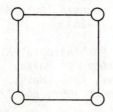

How many ways can this be done? (The symmetry group is D_4).

6. How many bead bracelets with 5 beads each can be made, if three colors of beads can be chosen: red, yellow and blue.

182

7. Assume that Army officers decide to paint the outside walls of the
    Pentagon building red, white and blue. Two paintings are
    considered equivalent if one is a rotation of the other; but,
    unlike the bracelet, the Pentagon cannot be flipped over. In how
    many different ways can this be done -- with the added provision
    that each color is to be used for at least one wall?

8. A baton is painted with n equal-sized cylindrical bands. Each
    band can be painted one of 3 colors. How many different colorings
    are possible?

9. How many different ways are there to 2-color the 64 squares of an
    8 x 8 chessboard that rotates freely?

10. Think of the set of all six-digit binary words -- that is, strings
    of length 6 composed of 0's and 1's. Ex: 010110. In some
    applications, two six digit words are considered equivalent if we
    get one from the other by applying the "cyclic permutation"

$$a_1 a_2 a_3 a_4 a_5 a_6 \longrightarrow a_6 a_1 a_2 a_3 a_4 a_5$$

    a number of times.

    Ex. 010110 is equivalent to 110010

    Find the number of such non-equivalent words.

11. Let G be a group of symmetries of the finite set X. Suppose that
    for all x and y in X there exists a g in G such that $y = gx$.

    Prove that $|X|$ divides $|G|$.

12. Allow 6 compartments and 4 colors in the Color Wheel problem. How
    many distinct color wheels are possible?

13. Determine the orders of the rotation groups of the following
    regular 3-dimensional solids.

    1) the tetrahedron (has 4 congruent triangular faces)

    2) the octahedron (8 triangular faces)

    3) the dodecahedron (12 pentagonal faces)

183

4) the icosahedron (20 triangular faces).

14. 1) Is every group of order n^2, n a positive integer, abelian?

    2) Is every group of order p^3, p a prime, abelian?

15. Suppose the group G has order p^n, p a prime and $n \geq 1$. Show that the center, Z(G), of the group consists of more than the identity.

16. Let Aut(X) designate the set of all permutations of the finite set X. Suppose G is a group of symmetries of X.

    1) For $g \in G$, show that the mapping $f_g: X \longrightarrow X$ by $f_g(x) = gx$ is an element of Aut(X).

    2) Show that the mapping $\pi : G \longrightarrow$ Aut(X) by $\pi(g) = f_g$ is a homomorphism of groups.